District Health Care

Challenges for Planning, Organisation and Evaluation in Developing Countries

SECOND EDITION

R. AMONOO-LARTSON MD FWACP MPH DCH DTM&H
Director, Health Services Consultancy Ltd. Box 1659, Tema, Ghana; formerly Deputy Director of Medical Services, Ministry of Health, Ghana

G. J. EBRAHIM FRCP (Edin & Glas) DCH (Lond)
Emeritus Professor of Tropical Child Health, Institute of Child Health, London; Editor of the *Journal of Tropical Paediatrics*; formerly Consultant in Child Health, Ministry of Health, Tanzania

H. J. LOVEL BSc PGCE MPhil
Senior Lecturer in Health Care Planning and Evaluation, Institute of Child Health, London

J. P. RANKEN BA MIPM LHA
Formerly Senior Lecturer, Institute of Child Health, London

MACMILLAN

© Copyright text R. Amonoo-Lartson, G.J. Ebrahim,
H.J. Lovel, J.P. Ranken 1984, 1994

All rights reserved. No reproduction, copy or transmission of
this publication may be made without written permission.

No paragraph of this publication may be reproduced, copied or
transmitted save with written permission or in accordance with
the provisions of the Copyright, Designs and Patents Act 1988,
or under the terms of any licence permitting limited copying issued
by the Copyright Licensing Agency, 90 Tottenham Court Road,
London W1P 9HE.

Any person who does any unauthorised act in relation to this
publication may be liable to criminal prosecution and civil
claims for damages.

First edition 1984
Reprinted once
First ELBS edition 1985
Reprinted four times
Second ELBS edition 1994
Second edition 1996

Published by MACMILLAN EDUCATION LTD
London and Basingstoke
*Associated companies and representatives in Accra, Banjul,
Cairo, Dar es Salaam, Delhi, Freetown, Gaborone, Harare,
Hong Kong, Johannesburg, Kampala, Lagos, Lahore, Lusaka,
Mexico City, Nairobi, São Paulo, Tokyo*

ISBN 0-333-57349-8

Printed in Hong Kong

A catalogue record for this book is available from the
British Library.

Contents

Preface vii

1 The Need for Management in District Health Care 1
  Who is this book for? In what district? 1
  Why is a new approach to health care needed? How have the old approaches failed? 4
    Health services continue to grow yet health problems and health service inadequacies persist 4
    Why no improvement in health? 9
    Additional problems exacerbating the difficulties in planning 12
  A new approach to health care and its management requirements 13
    The origins of the new approach 13
    Features of primary health care 13
    Management requirements for successful primary health care 14
    How have countries responded to the concept of primary health care? 24
  Doctors as managers and leaders 26
    Functions of leadership 26
    Decision-making 27
    Key abilities of a good leader 27
    What are the desirable traits of a successful leader? 29
    Factors influencing the work of the district manager 29
    Management styles 30
  Further Reading 32

2 Finding Out About Health Needs in the District 33
  Who? What? Where? When? Why? in ill health 34
    Which age groups contain most people and which age group is increasing fastest? 34
    Who gets sick? Who dies? 34
    Who needs maternity care? 35

What can be done about high maternal mortality?	36
Requirements for safer motherhood	38
What are the health problems?	38
Where are the health problems in the district?	46
When does ill-health occur?	48
Rapid epidemiological assessment	48
Why does ill-health occur?	52
What is wrong with the existing health services?	56
Are the services coping?	56
Coverage	57
Are people utilising the services?	59
Is the 'at-risk' concept being used in provision of health services?	60
Is there adequate quality of care?	61
Is staff morale high?	62
Do staff interact?	62
Is there regular health services evaluation?	63
Identifying local resources	64
Who is providing health care? Who do people go to for advice? Where? When? At what cost?	64
People other than health workers as resources	72
Resources of material and labour	74
Financial resources	75
Natural resources	77
Methods for finding out what is happening in the district	78
Further Reading	82

3 Making a Health Plan for the District — 84

What is a plan?	86
Future	87
Goals	87
The planning process	89
The health planning and implementation cycle	89
Dangers of planning and why planning sometimes fails	89
Planning is a learning process	93
Key concepts in effective district health planning	93
Resource allocation and budgeting	114
Examples of different elements of a district health plan	117
Writing project proposals	132
Project formulation	133
Getting feedback	133
Writing the detailed project proposal	134
Anatomy of the proposal paper	134
Further Reading	139

Contents

4 Building the Health Organisation in the District ... 140
The health organisation as an internal system within an external environment ... 141
Formal and informal organisation ... 144
An organisation as a skill pyramid ... 146
An organisation as a network of individuals ... 146
An organisation as a system or series of systems for getting things done ... 146
Organisational culture ... 151
Some key principles in an effective organisation ... 151
The key elements of an effective organisation ... 155
   What it means to manage a district health organisation ... 157
   Managing within the local socio-cultural environment ... 160
   Is the district health organisation functioning well? ... 162
   Organisational change ... 163
   The levers of change ... 166
Further Reading ... 167

5 Practical Management: Putting Plans into Action ... 168
Management by objectives ... 171
Participatory management ... 176
Standards ... 178
Safety standards ... 180
Personal skills of the manager ... 183
   Managing time ... 183
   Delegation ... 187
Teamwork ... 189
   What makes a good team? ... 193
   Group work ... 195
   Meetings ... 197
Motivation ... 199
Communication ... 208
   Problems with communication ... 209
   The rules of good communication ... 209
   Direction of communication ... 210
   The psychology of communication ... 210
Management of change ... 212
Introducing change ... 215
   Recognising the need for change ... 215
   Planning the change ... 215
   Getting agreement ... 216
   Implementing the change ... 217
   Checking and monitoring change ... 218
   Checklist for organisational change ... 222

Managing conflicts	223
The problem-solving approach	223
Bargaining	225
Use of 'third parties'	225
Avoiding or ignoring the conflict	226
Confrontation	226
Giving support to supervisors	227
The job of supervisors	227
Support for supervisors	228
Personnel – the management of health workers	229
Manpower planning	230
Training	232
Staff development	234
Maintaining standards and discipline	236
Counselling – or helping staff with their problems	241
Finance	242
Sources of finance	242
Operating budgets	244
Costing information	245
Cost-benefit analysis	246
Financial reports	247
Economic recession and district health financing	247
Buildings	248
Supplies and stores	250
Drugs, vaccines and other pharmaceuticals	252
Vaccines	253
Transport	253
Further Reading	255
6 Getting Feedback: Monitoring and Evaluation	256
Why do we need feedback?	256
What is monitoring?	258
Methods of monitoring	259
What should be evaluated?	260
What needs to be evaluated depends on the community diagnosis of problems and resources	260
What needs to be evaluated depends on the plan of action for the health team in the district	261
What are the key components of evaluation which often get neglected?	261
What are the essential elements of primary health care which need to be evaluated?	263
Which level? Which component? – levels of evaluation	266
What level of feedback and which component in a programme?	266

	Which specific questions can feedback from a health care programme consider?	267
	How can data be obtained to find out what is going on?	267
	Quick or long? and what disciplines in getting feedback on a programme?	268
	Examples of data collection forms and systems used in local communities	271
	Choice of evaluation method	272
	Experiment design	272
	Quasi-experimental design	272
	Health services research (HSR)	273
	Routine health data	275
	Who is to obtain the feedback information?	277
	Where should the feedback be done?	278
	When should feedback information be obtained?	278
	Constraints on getting feedback	279
	Further Reading	280
7	**Future Prospects: Challenges for Change**	281
	Management issues likely to arise during the next decade	286
	Obstacles and constraints	288
	Opportunities for further growth	289
	The untapped resources	291
	Beyond primary health care?	292
Index		293

Preface

Following the Alma-Ata declaration on Primary Health Care (PHC) as countries began to reorient the health services to achieve the goals of availability, accessibility and affordability of health care for all citizens, a number of management issues came to the forefront. The publication of the first edition of *District Health Care* was timely. Together with its sister title *Paediatric Practice in Developing Countries* which describes the technical aspects of PHC, the first edition of *District Health Care* provided the necessary conceptual framework for the planning and management of PHC. A number of colleagues around the world have since written to say how helpful the two titles have been, especially for training.

The 1980s witnessed a remarkable improvement in the global health situation. The Child Survival Revolution helped to focus attention on simple technologies which, when universally applied, have proved to be immensely successful. Encouraged by these successes the world leaders have outlined the major challenges for the 1990s at the World Summit for Children. If the momentum is to be maintained then one overriding requirement would be for appropriately trained health personnel. It is not good enough merely to tag on PHC as an addendum to a curriculum already overburdened with medical rarities. PHC has to be at the heart of the training of all health workers. Radical and imaginative revisions of existing training programmes are needed.

Progress in the 1980s was achieved in a climate of global economic recession, which has continued into the 1990s. Countries of the developing world have been additionally hit because of the punitive structural adjustment policies imposed on them by the World Bank and the International Monetary Fund. The challenge facing the developing world at present is for planning to consolidate the progress of the 1980s and build on it in a climate of continuing economic difficulties.

Management sciences have experienced marked new developments during the last two decades. Technological advances have changed the systems of industrial production and commerce as well as the characteristics of the work force. New techniques in management have followed in the wake of these changes. Many such new developments stem from a growing convergence between behavioural as well as social sciences and the discipline of management. Several of these conceptual changes have been referred to in this second edition, and a number of new management techniques have been described.

Changes are also taking place in the health systems of the more developed countries. Partly to contain costs and partly to effect an equitable distribution of resources, health service reforms are occurring in most countries. Underlying these reforms is the drive to transfer resources from care in the hospital to care in the community. The result of the reforms has been the increasingly important role of managers. This is understandable. The annual budget of an average District health programme could well exceed that of a medium sized commercial or industrial concern. It is not surprising that efficient management of health resources – material, financial and human – is coming under close scrutiny. Efficiency requires the targeting of resources on key result areas. This has meant an increasing convergence between health management and the epidemiological sciences. New initiatives in training are called for in order to enable health personnel to understand management and for managers to understand epidemiological and health care issues. Several of these matters are addressed in the final chapter of the book.

We hope that like its predecessor this second edition would provide the conceptual framework for health planning towards the goal of Health For All – 2000.

G. J. Ebrahim

1 The Need for Management in District Health Care

WHO IS THIS BOOK FOR? IN WHAT DISTRICT

People working at District level in many countries are now frequently faced with the task of putting a national Primary Health Care policy into action. Sometimes detailed plans exist as, for example, in Ghana, the Sudan or India; but in many other countries there is no detailed plan. Many problems are being experienced and the object of this book is to identify the ways in which some of these problems can be tackled. The book is intended for anyone who can identify with any of the issues in table 1.1.

Frequently people experiencing such problems will be members of a District Health Team (DHT). Often some of them will have had a clinical training (for example, as a doctor, a nurse or a nutritionist) and be faced with much non-

Table 1.1 **This book is intended for anyone who can identify with any of the following issues.**
(Tick if they apply to you.)

		Tick
1	'I'm too busy, I never have time'	
2	'We can't do it, there aren't enough midwives'	
3	'There aren't enough drugs'	
4	'There's no money left to pay transport costs'	
5	'It's the fault of the bureaucrats'	
6	'I never knew those people had such problems, they never used our services'	
7	'We have done so much to expand the services'	
8	'I thought our job was just inside the hospital'	
9	'There are no people to control all this rubbish'	
10	'We're a new team, we don't know what to do'	
11	'People don't recognise the work we do'	
12	'What could traditional healers do?'	
13	'What could families and the local community do? – they are not trained'	
14	'I don't know how to do evaluation'	

clinical work concerned with planning, organising and supporting different aspects of an 'outreaching' District health care programme. Environmental health personnel may also be facing difficulties because they are highly trained in some areas (for example, maintenance of communal food markets) yet may find themselves doing a job where this skill is needed only very occasionally. The rest of the time they find that they are being asked to provide refuse or sewage services in a vast area with limited personnel and resources. Increasingly, in several countries competent and trained lay health administrators are becoming available. Their special skills can contribute greatly to the effective management of rural health services.

Districts vary from one country to another but from an organisational point of view many have similar features. In this book a 'model' for provision of health care is described which is already widely used in a number of African and Asian countries and is increasingly being adapted for use elsewhere. The principles are relevant everywhere. For example, there is increasing recognition in all countries that as much caring as possible should take place not in hospitals, but at the Primary Health Care level, within the community. For instance, in some rural parts of the United States traditional birth attendants do many deliveries, and physicians' assistants carry out many services in the absence of university-trained doctors. Increasingly, people's views about their services are being taken more into account.

The average District in mind has a population of between 200 000–500 000 though it could be as low as 100 000 in sparsely-populated countries or as high as one million. It covers varying areas from one country to another. There will be a District General Hospital (and possibly other hospitals) and a number of Health Centres and health posts, on the general principle of one Health Centre for every 100 000 population and one sub-centre for every 10 000 people.

All nations are now actively working towards the establishment of a viable Primary Health Care programme. The District Health Team is expected to face this challenge and make the programme function effectively. The aim of this book is to look at the reality of translating plans into action and concepts into practice. Table 1.2 lists some of the problems discussed in the book.

Table 1.2 **Do you experience these problems? (Tick)**

Problems at the District level

No agreed plan for the District
No defined targets
Lack of adequate data to define District problems
Ignorance of activities taking place in the District
Lack of financial planning and a systematic way of allocating resources
Lack of interest, enthusiasm and understanding of what has to be done
Under-worked people
Over-worked people

Table 1.2   (contd.)

Problems in Hospitals

Wrong kinds of services are provided
Overwhelmed by demands
Divorced from community
No assessment of the community's health needs
Poor utilisation of resources, e.g. nurses' skills
No plans or targets
No clear policies or delegation
Hospital activities not integrated with Primary Health Care
Unrealistic aspirations
Poor financial control
No evaluation of hospital services

Problems at Health Centre level

Poorly trained staff
No regular in-service training
Lack of contact with villages
No support for community work
No procedures for routine activities
No defined targets
Lack of initiative
No community diagnosis
Over-strict demarcation of duties
Isolation from District officials and other workers
Unreliable deliveries of drugs and supplies; delayed payment of wages etc.
Poor morale
Lack of interest
Overspent budgets
Malpractices such as absenteeism, pilfering of drugs and misuse of vehicles
Unused and broken-down equipment
Badly-used buildings
Shortages of drugs, supplies and transport
Poor career structure and lack of job satisfaction
Problems at Health Centre level not recognised by District

Problems in the community

No services within reasonable reach
Mothers and babies die of preventable illness
Poor hygiene
Poor water supply
Lack of knowledge; ignorance
Shortage of leaders
Apathy, sense of futility
No priorities for tasks to be done
No links between health – education – agriculture – community development, i.e.
   no interdepartmental co-ordination

Table 1.2 (*contd.*)

Over-all problems

Lack of interest in management
Poor understanding of functions and roles, and of the health problems of the community
Misuse of resources
Lack of commitment and misplaced enthusiasm
Little contact with other departments like agriculture, education, social welfare, community development etc.
Constantly changing policies and priorities.

Problems such as these do not apply in all places and frequently progress is made in very adverse conditions. Usually the best solutions to problems are those which people work out for themselves and which best fit the local situation. This book attempts to provide some guidance to help managers see more clearly the nature of some of their problems and to suggest ideas which may be helpful in resolving them.

WHY IS A NEW APPROACH TO HEALTH CARE NEEDED? HOW HAVE THE OLD APPROACHES FAILED?

Health services continue to grow yet health problems and health service inadequacies persist

Despite a heavy infusion of health care resources in many countries in the last 20 years, there is still a lack of health care in most areas. Table 1.3 shows data from Ghana where there were huge increases in the number of health personnel between 1960 and 1975, so that by 1975 the health worker: population ratio looked quite good by any standard.

This table shows that since 1960 there have been remarkable increases in the resources infused into the health services of Ghana, most notably in terms of manpower and facilities. Yet despite this heavy infusion of resources, the government recognised that by 1975 there had been little or no impact on the health status of the population in general, and in particular on the 70 per cent of the people who live in rural areas. Similar experience is not uncommon in several countries.

Even with the increase in health manpower and hospital beds, clearly the health problems and health service inadequacies are continuing. Table 1.4 shows the distribution of doctors in Ghana in 1975. Over a third were in the capital city, Accra; another third were in the large towns and only the

Table 1.3 **The growth of health manpower and hospital beds, 1960-1975, Ghana**

Category of health personnel	1960	1975	Per cent increase 15 years	Population per health professional 1960	1975
Physicians	383	1 031	169%	17 564	9 625
Dental surgeons	19	60	216%	354 052	165 383
Nurses	1 554	6 153	296%	4 329	1 613
Midwives	130	4 932	3 694%	51 746	2 012
Hospital beds and cots	5 787	12 973	124%	1 162	765

Source: 1960 figures, The Health Services in Ghana, (1961), D. Brachott; 1975 figures, National Health Planning Unit, 1975 population estimate 9 923 000.

Table 1.4 **The problem of the distribution of doctors (Ghana, 1975)**

	% doctors	% population
Accra (the capital city)	34	7
Towns of more than 20 000 population	33	11
Places of less than 20 000 population	33	82

Source: Ghana Ministry of Health (1977), *Health Data book*, National Health Planning Unit. Population data from 1970 Census.

remaining third were to be found in the smaller communities, of under 20 000 people, where the majority of Ghana's population live. A similar pattern of distribution was likely to be found for midwives and environmental health workers. Table 1.5 gives the pattern of distribution of doctors in several other developing countries.

Apart from this obvious mismatch between health care provision and population distribution, another problem is apparent. Many deaths in the developing world still occur for reasons which could have been avoided through the provision of simple appropriate care. In many parts of the developing world up to 40 per cent of all children die before they reach school age (see table 1.6).

The tragedy is that, as in so many countries, the diseases could be prevented or controlled with Primary Health Care, by the use of immunisation, simple medications, environmental alterations and health education for the people. In the past ten years virtually no impact has been made on the prevalence of communicable diseases in many countries and some, like yaws and cholera,

Table 1.5 **Distribution of doctors between the capital and the rest of the country, 1968**

Country	Population : Doctor Ratio		
	Nationwide	Capital city	Rest of the country
Kenya	10 000	672	25 600
Thailand	7 000	800	25 000
Guatemala	4 860	875	22 600
Jamaica	2 280	840	5 510
Pakistan	7 400	3 700	24 200
Philippines	3 900	1 500	10 000

Table 1.6 **Measures of ill-health in developing countries, 1970–75**

Region	Life expectancy at birth (years)	Infant mortality per 1000 live births	Mortality aged 1–4 per 1000	Crude birth rate per 1000
Tropical Africa	41	200	40	22
Northern Africa	52	150	26	15
Western South Asia	54	135	22	14
Middle South Asia	48	145	25	17
Eastern South Asia	51	120	18	15
Melanesia	48	150	–	17
East Asia	61	70	7	10
Caribbean	63	64	7	9
Tropical South America	61	100	10	9
Middle America	62	70	9	9

have increased dramatically. In addition, the maternal mortality rate remains high, for example at over 40 per 10 000 deliveries in the rural areas of Ghana and 42 per 10 000 births in India, which is several times more than the current rate of less than 2 per 10 000 in Europe.

Not only are there high death rates due to preventable diseases but also there are high rates of blindness, lameness and other forms of disability. Many of these could be prevented or minimised with a Primary Health Care programme. A few countries have begun to measure the effects of high morbidity rates on the national economy. For example, the National Health Planning Unit in Ghana conducted a technical analysis of disease problems and their impact on the level of health of the people. This is measured by estimating the number of days of healthy life lost due to sickness, disability or death caused by each sickness. The following are the sixteen major causes of sickness, disability and death found (see table 1.7).

Table 1.7 **Disease problems of Ghana ranked in order of impact on health status**

Rank order	Disease classification	Days of healthy life lost*	Per cent of total
1	Malaria	58 427	15.4
2	Prematurity	34 432	9.1
3	Measles	23 033	6.1
4	Birth injury	22 612	6.0
5	Sickle-cell disease	21 797	5.9
6	Pneumonia, child	20 857	5.5
7	Kwashiorkor, marasmus	19 312	5.1
8	Dysentery & gastro-enteritis	17 022	4.5
9	Neonatal tetanus	14 047	3.7
10	Accidents (all kinds)	11 137	2.9
11	Tuberculosis	10 097	2.7
12	Cerebrovascular accidents (stroke)	8 915	2.4
13	Pneumonia, adult	8 743	2.3
14	Psychiatric disorders	8 542	2.3
15	Cancer	7 315	1.9
16	Pregnancy, complications of	6 005	1.6
Total of these 16 diseases		292 293	77.4%

* *Days of Healthy Life lost due to sickness, disability and death*: This is the number of days of healthy life estimated to be lost due to onset of disease (including the number of days in future years lost due to premature death or disability) in one year in population of 1000.

Example:
A man of 26 years develops tuberculosis and after 4 years sickness causing partial disability (25% disabled) he dies at age 30 years. He has lost 34 years of future expected life (plus one calculated year of 25% ($\frac{1}{4}$) disability over his 4 years with the disease; $4 \times \frac{1}{4} = 1$ year) i.e. 35 years of 365 days = 12 755 days of healthy life lost.

With the incidence of tuberculosis in Ghana being 2 per 1000 population per year, the average illness duration being 5 years and the average disability being 25%, as in the case above, in a population of 1000 two people will get tuberculosis disease each year. The case fatality rate is 30% (not everyone with tuberculosis dies, just 3 in 10 people). So the number of days of healthy life lost due to tuberculosis in a population of 1000 is 10 097 days per year.

Regional variations in mortality experience occur in all countries including the more developed countries. In the less developed countries, however, there is also the glaring urban/rural disparity so that infant and child mortality rates in rural areas are about a third higher than those for the cities. In recent years a further twist has been added to the question of disparities in the form of growing numbers of squatter settlements in most large cities. The mortality experience of the urban poor often exceeds that of the villages in spite of the relative proximity of the former to health care facilities. For example, in Manila the infant mortality rate is three times higher in the squatter areas than in the rest of the city, and in the *bustees* of New Delhi the over-all child mortality rate (0–5 years) reaches 440 per 1000 amongst certain groups.

Unequal distribution of health facilities resulting in disparities of health care is usually due to political and socio-economic factors. But disparities also indicate poor planning and management of health resources. Similar forces also operate at the Regional and District level as at the National level, leading to differences in health experience between communities. Some common symptoms of poor health management and planning are described in table 1.8.

Table 1.8 **Some symptoms of poor health management**

1. Adverse mortality and morbidity from preventable illness. High infant mortality. Major differences in mortality rates and life expectancy between communities.
2. Health facilities geographically inaccessible on account of urban bias, and rural inadequacy. In some countries up to 85% of the population in rural areas have no access to any form of modern health care.
3. People cannot afford to pay for health services, or the costs of time away from work, transport, etc.
4. Curative care is emphasised to the neglect of prevention and early treatment. Institutional care is the common practice rather than care provided near the home within the community.
5. More money is spent on hospitals than on simple Primary Health Care facilities, with up to 80% of health spending being on 'Western' type hospital care.
6. The wrong kinds of health workers are trained, with an emphasis on training highly qualified doctors and specialists instead of auxiliaries with basic skills.
7. Health services are often imposed from above and do not always have the support of local communities.
8. The Health Care provided is inappropriate for the main health needs. Infections, parasitic and respiratory diseases, are often widespread and have relatively simple preventive solutions if implemented on a wide scale, but health resources are often devoted mainly to curative services.
9. Demographic pressures are increasing with disproportionate increases in numbers of babies and children. This causes demands for maternal and child health services which are neglected in favour of fashionable intensive care.
10. Severe economic problems force governments and planners to impose drastic cuts so that essential drugs and supplies become scarce with periodic shortages.
11. Widespread poverty causes environmental problems, shortage of water and natural resources.
12. Breakdowns in conventional medical systems where they exist, e.g., overcrowding of facilities, misuse of technical equipment, shortages of supplies, unresponsiveness to chronic diseases and handicap.
13. Isolation and weak support of existing primary health care services and workers, leading to a discrediting of the contribution they can make.
14. Difficulties in implementing individual 'vertical' health programmes on a massive scale (e.g., malaria prevention, leprosy control, programmes for eye diseases) and maintaining their effectiveness on a permanent basis.
15. More interest amongst some health planners in detailed research and lengthy reports than in contact with local people and implementation of appropriate programmes.
16. Unwillingness of the health establishment to involve existing human resources in the community for extending health coverage.

Why no improvement in health?

The data presented in the preceding paragraphs indicate that despite a very great increase in the resources devoted to the health sector in the last ten years, there has been little improvement in over-all health status and health service provision for the majority of people. One basic reason for this seems to be that the health services have been doing the wrong things because of misplaced priorities. This is as true at the District level as at the National level. The existing health service system funnels resources towards the minority of populations having access to hospital-based services catering largely for specialised health problems (see figure 1.1).

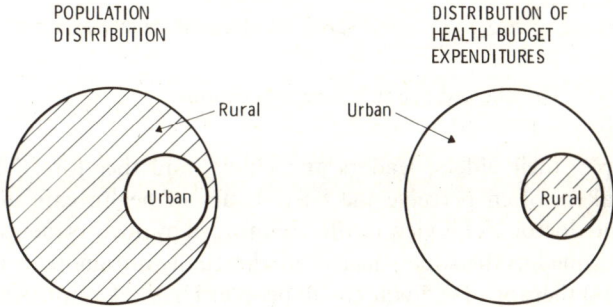

Figure 1.1   Undesirable distribution of health budget expenditures

In many countries, for understandable but self-defeating reasons, hospital-based curative services requiring highly trained personnel and expensive equipment have received the greatest amount of attention and have commanded the bulk of resources (up to 80 per cent) set aside for health care by the government. This has created a paradoxical situation. Because of inadequate public health and Primary Health Care services, patients with preventable conditions have over-loaded the hospital services so that tertiary care facilities are being used for primary care. On the other hand, overcrowding in hospitals leads to greater demand for more hospitals. As more resources are put into the construction and equipping of hospitals and the training of sophisticated health workers required for their operation, even less resources become available to develop the Primary Health Care system (see figure 1.2).

For most preventable conditions (parasitic diseases, nutritional disorders, common infections, illnesses of childhood, etc.) the elaborate hospital care required after the disease is established is not only expensive, but is also only partially effective, thus compounding the consequences of this inappropriate allocation of resources. It is an unfortunate paradox that the *demand* for services occurs only after illness becomes evident: while the *need* for services is long before illness occurs.

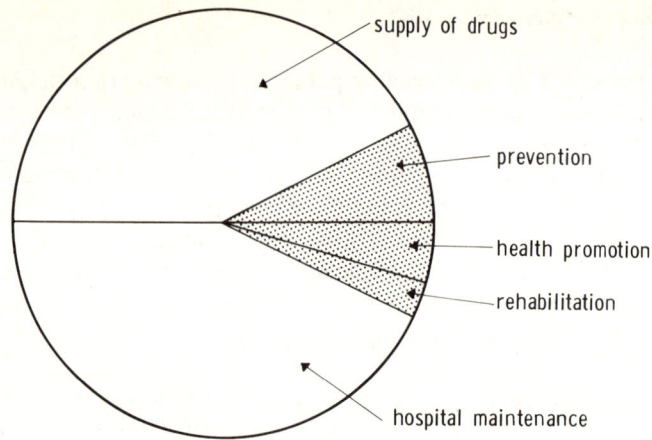

Figure 1.2   Unbalanced health budget allocations

The people, their elders, leaders and chiefs, and therefore the political decision-makers often perceive the basic issues of health care in terms of hospitals and doctors. This view is often reinforced by doctors themselves who have been trained in the sophisticated, intellectually intriguing disciplines of diagnosis and therapy of individually ill people. Their attention is focused on the sick who seek their help. But the need is to maintain the health of *those who are not yet ill*.

This situation is strikingly illustrated in figure 1.3 describing the health care dilemma of the average developing country, taking Ghana as an example. It shows how the financial resources of the nation are being allocated in reverse proportion to the number of people in need!

Five main reasons have been identified as the cause of this situation in which many countries, developing and developed alike, find themselves. These are:

(1) Focus on the construction of facilities rather than the provision of services.
(2) Over-sophisticated training which takes place largely inside hospitals with emphasis on specialised hospital-based services rather than preventive and promotive services.
(3) Poor and unequitable deployment of health staff.
(4) A 'top-down' health care delivery system with a noticeable lack of co-ordination with other sectors (social welfare and community development, water and sewerage, education, agriculture, etc.) and little or no community involvement.
(5) Unbalanced health budget allocations.

The present health care system of many countries can be likened to a pyramid with the Teaching or District Hospital at the top, as the case may be,

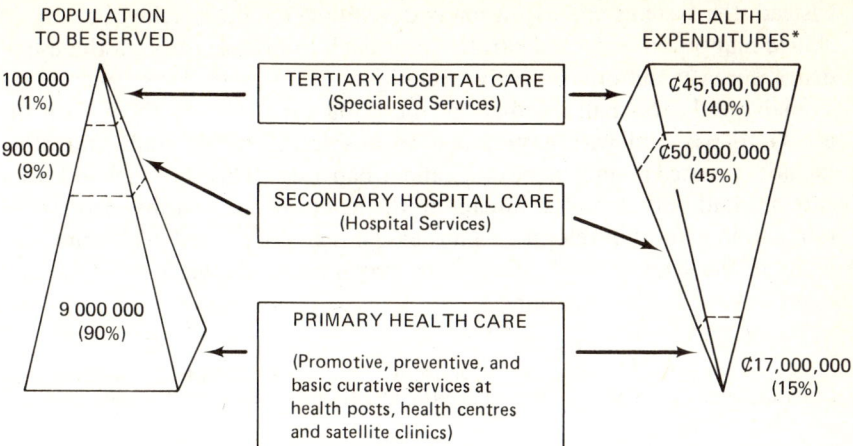

*Approximations based on 1975/76 estimates
₵ = Cedi, the currency in Ghana

Figure 1.3　　The health care dilemma in Ghana. (The distribution of funds and personnel for primary health care compared to costly hospital-based care is in inverse proportion to the numbers of people that need to be reached. The Health Care Pyramid is upside down!)

　　　　　　　Source: Primary Health Care Strategy for Ghana, National Health Planning Unit, 1979.

and a network of health posts and dressing stations at the bottom. This is a system based on health units (that is, structural facilities). It focuses attention on 'bricks and mortar' rather than on health services. This emphasis on structural facilities creates false 'needs' among the people for more facilities. Good health becomes synonymous with the provision of a doctor and a hospital rather than the enjoyment of a disease-free environment.

Under these conditions, each community without a health post believes it should have one for reasons of prestige; each community with a health post wants to expand it into a Health Centre; and each community with a Health Centre sees a 'need' to add a surgical theatre and ward and bring a surgeon to town. The pressure continues upward to Regional and National levels with its attendant burgeoning demands for financial and manpower resources. This upward pressure absorbs all available resources and leaves an ever-increasing vacuum at the bottom of the pyramid where the real health needs of the nation lie.

The approach adopted in many countries up until the 1960s and 1970s was based on systems of health care in use in Western developed countries, forgetting that the main strength of the health system (for example, in Britain) lay in the grass-roots viz. the general practitioner and the health visitor.

Instead, the systems created in many developing countries relied heavily on skilled manpower and on a curative approach to disease through the use of drugs, surgical and other invasive techniques. But in such a system care was available only to small numbers of the population, mainly those with the ability to pay and with easy access to hospitals, trained staff, and other facilities situated mainly in the cities and urban areas. This approach to health care has had only a modest impact on the health of the vast majority of the population in most developing countries. Though not present in all countries, many of the symptoms of poor health management shown in table 1.8 and figure 1.4 are frequently found.

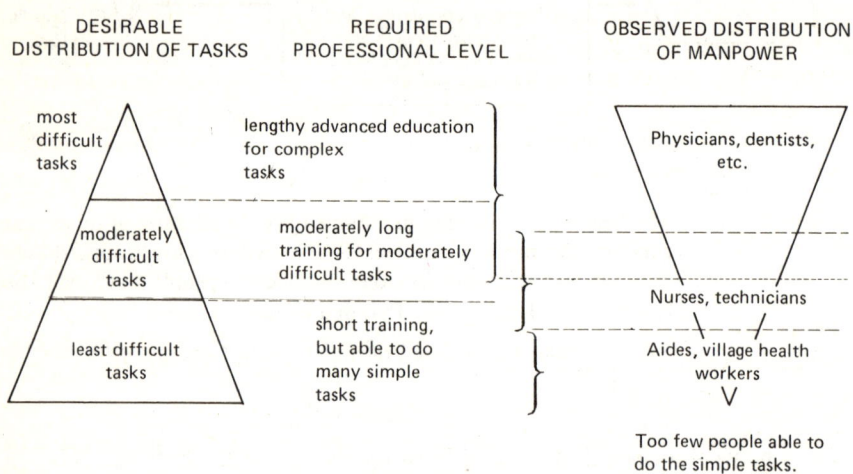

Figure 1.4   Why no improvement in health? Poor deployment of health staff

Additional problems exacerbating the difficulties in planning

A number of interrelated factors compound the problems of misplaced priorities:

(1) Population growth and its impact on the per capita expenditure available for health and development.
(2) High inflation together with the devaluation of many currencies.
(3) Economic recession with heavy debt burden leading to the imposition of structural adjustment policies by the World Bank and the International Monetary Fund.
(4) The spreading epidemic of HIV (Human Immunodeficiency Virus) infection causing heavy strain on national economies, and the virtual breakdown of health services.

(5) Degradation of environment with resulting climate change; other natural disasters, wars, insurgencies and political unrest.

The crude birth rate in many countries is currently between 25 and 45 per 1000 population. At the present rate of population growth of between 2 and 3 per cent per year the population of many countries will double in 25 to 35 years. Such growth in population means that each year more and more people must share the 'national cake' of the health services.

Even when governments can increase health expenditure the amount of goods and services the money will buy may still become progressively smaller with inflation. The effects of inflation are further compounded by the debt crisis, particularly in Latin America and sub-Saharan Africa. For example, in the 1980s more than 25 per cent of export earnings in sub-Saharan Africa, and more than 40 per cent in the case of Latin America/Caribbean went into debt servicing. The ensuing programmes of structural adjustment have resulted in cutbacks in government expenditures on health, education, social services, and in food subsidies.

Health services have been put under further strain by the spreading epidemic of HIV infection, so that in some countries up to half the hospital beds are taken up by patients suffering from the Acquired Immunodeficiency Syndrome (AIDS).

These reasons explain why the District Health Teams in many countries should not expect to see real improvements in the District Health Budget in the foreseeable future.

A NEW APPROACH TO HEALTH CARE AND ITS MANAGEMENT REQUIREMENTS

The origins of the new approach

In the light of these problems concerning the existing health care system, many countries are now evolving more effective ways of improving health in their communities. Some of the significant developments are summarised in table 1.9.

Features of Primary Health Care

A number of countries are adopting new approaches to health care. There are two major objectives, (1) to design health services which can reach the majority of the people, and (2) to prevent and treat the preventable disease problems which are presently responsible for much ill health (morbidity as well as disability) and mortality. Both these objectives are to be achieved through community involvement and participation which require a high level of awareness.

The principal means of implementing this approach is through a Primary Health Care system aimed at reducing the rates of mortality and morbidity caused by conditions for which prevention, easy treatment and control exist. Prominent among these causes in many countries are communicable diseases, nutritional deficiencies and manageable complications of pregnancy. Priority for health improvement is then determined by which procedures can produce the greatest reduction in the unnecessary burden of sickness, disability and death, at the least cost. Most of the health improvement procedures that will have the greatest impact are simple tasks that can be carried out by people without sophisticated training using simple equipment, provided there are satisfactory organisational arrangements for the needed support and supervision. This new approach to health care provision is seen as part of the national effort in social and economic development with community involvement at each level and support from health centres and hospitals. Together they make up the system for Primary Health Care.

However, there is no 'right' or 'wrong' way of providing Primary Health Care in a country, or even to all regions of a single country. Countries and communities vary in terms of size, geography, climate, population, communications, level of political, economic and social development, health needs and resources, and local leadership. Systems of providing health care need to be evolved which meet each locality's circumstances and problems. It is important also that a country feels confident that it has developed a system which meets its own needs, and which, because it has been developed by its own people, has the commitment and determination of its people to make it work. But there are certain features of primary health services which are increasingly being widely adopted. These are shown in table 1.10.

Management requirements for successful Primary Health Care

The success of a Primary Health Care system depends on a number of factors, in particular National and local commitment, National strategy to provide guidelines, a local District plan building on local experience, a District Health Team as part of a multi-level structure with clearly defined roles which recognise the different tasks in hospital and community, and local community involvement. The following requirements for success in setting up a District programme for Primary Health Care can be identified:

(1) *Awareness* among health workers, health planners and lay people of basic health problems.

Table 1.9 **Significant new developments to solve the problems in existing health care provision**

1. A commitment on the part of a number of governments to give priority to primary health care for disadvantaged groups.
2. Recognition that many diseases widespread in the population lend themselves to relatively simple non-medical solutions, e.g., improvements in water supply, hygiene, nutrition, etc.
3. Recognition of the social nature of much ill-health like the lack of services for the rural areas and for the urban poor which calls for a sociological and political rather than medical approach.
4. Recognition of community development, whereby the development of a community in all its aspects, family welfare, farming, income-earning activities, social functioning, child-rearing, community action, etc., is seen as having a vital part to play in improving not only health, but other aspects of development.
5. Greater awareness of the need for co-operation between agencies, e.g., education, agriculture, health, to ensure a sharing of resources and a co-ordinated approach to problems.
6. Development of 'Bottom-up' as opposed to Top-down' planning, i.e., taking the needs, resources and opportunities in local communities as the starting point for planning health services, as opposed to planning on the basis solely of needs and policies as seen at the national level.
7. Increasing awareness amongst ordinary people of their lack of health care and growing demands for something to be done about it.
8. Because of difficulties in implementing individual 'vertical' health programmes over unduly long periods (e.g., malaria, leprosy) a recognition of the need for 'general care' rather than 'specialist cure' programmes.
9. Recognition of the value of 'screening' techniques for large numbers especially among the vulnerable groups like children and pregnant and lactating women in the population to detect signs of disease at an early stage.
10. Recognition that with alternative methods of health care delivery, considerable improvements in people's health can be made at relatively low cost.
11. Recognition of non-financial resources which are often present in rural communities, e.g., social cohesion, traditional skills, enthusiasm when encouraged and motivated.
12. Growth of experience from different countries – e.g., of 'barefoot doctors' in China; health services 'where there is no doctor', in Mexico; planned health services at low cost in Tanzania.
13. Renewed interest by professionals in traditional cures and herbal medicines, in many parts of the world – in Africa, Asia, North America, China.
14. Application of 'appropriate technology' to health care, e.g., coloured strip for measuring arm circumference, rehydration spoons, simple weighing machines, low cost hand pumps for water supply etc.
15. Changes in medical training with a better understanding of psychological and social factors affecting health, and an appreciation of the need to train larger numbers of health workers with more appropriate skills.
16. Recognition of the enormous contribution which can be made to the provision of health care by groups hitherto under-utilised in the community, e.g., women, nurses, children.

Table 1.10 **Features of primary health care increasingly being adopted**

1 *Focus on the community*
   1.1 Emphasis on health care at the village or community level
   1.2 Use of local community workers at the first level of health care, drawn from and supported by the community
   1.3 Involvement of the community in the planning and running of their own health services
   1.4 Use of traditional methods and resources, e.g., the traditional birth attendant.

2 *Emphasis on hygiene and prevention of disease*
   2.1 Promotion of mother and child health services including immunisation and nutrition
   2.2 Environmental and public health given equal stress as curative care
   2.3 Emphasis on health education.

3 *Planning for services relevant to local needs*
   3.1 Identification of major health problems and adoption of specific programmes to combat them
   3.2 Regular evaluation to ensure continual improvement of health programme
   3.3 Strategies for improving coverage and for developing low-cost technology, e.g., limited drug formularies, low-cost waste disposal systems etc.
   3.4 Integration of health with other aspects of development, e.g., agriculture, education, community development and so on.

4 *Organisation of services to improve utilisation and quality of care*
   4.1 Provision of basic health care facilities within walking distance for the majority of the population
   4.2 A hierarchy of levels of health care comprising 3 key elements: local community; Health Centre/sub-centre complex; District Hospital
   4.3 Different levels of District health facilities becoming supportive of each other and of community-based health activities.

5 *Training based on locally assessed needs*
   5.1 Health workers at each hierarchy of care trained in preventive and environmental health as well as to diagnose and treat a limited number of common illnesses, less common diseases not included in the training being referred to the next higher level of care
   5.2 Continuing education to improve skills at all levels of care.

(2) *Commitment* to improve the health of the population, both for its own sake and as a foundation for economic and social development. This is as true at the District level as at the National level. This commitment needs to be expressed throughout the health system as a commitment to *Primary Health Care*. A balance needs to be maintained between primary and secondary care, and this may mean drastic and determined action where hospitals have expanded disproportionately. This entails a commitment to equitable *financing* and shifting of expenditure from the

tertiary level to the primary care level. Commitment is also needed amongst medical, nursing and other professional staff. Much is being done in some medical schools to select and train doctors to work in Primary Health Care where the attitudes, skills and conditions of work are very different from those appropriate to hospitals or private urban practice. A positive commitment to Primary Health Care is even more important at higher ministerial levels where policies are made and resources allocated for health care.

(3) The national commitment ideally finds expression in a *National Strategy*. Many countries have produced a National Health Plan which sets out a broad strategy and takes into account those factors which are best planned for nationally, for example:

(a) manpower planning to establish in broad terms the numbers and roles of health workers needed;

(b) training programmes, for example, medical assistants, public health nurses, environmental health workers;

(c) 'vertical' health programmes, for example, malaria eradication; simple and appropriate agreements on drug formularies and purchasing of drugs;

(d) design of effective health education material.

Within the framework of such a National Plan, effective District Plans need to be produced and attention given to the demanding task of implementing the plan.

(4) It is unlikely that a District Health Team will set about its task completely from scratch. Every society in order to survive has developed its own ways of coping with disease, and maintaining a state of health amongst its population. This will include feeding habits, housing arrangements, practices affecting pregnancy, child-rearing, care of old people, means of coping with disease, accidents and tragedy. Where traditional practices are effective they need to be maintained and encouraged. Health problems often arise because traditional practices, for example, birth-spacing, breast feeding, balanced diets, have been neglected in the face of 'modern developments' associated with such things as higher cash income, increased mobility, 'status' foods like carbonated drinks, powdered milk, packaged weaning foods etc. Beneficial practices and traditional health resources need to be encouraged and incorporated into the District Health Programme.

Some health programmes may already exist like Country Health Programmes organised on a national basis; health-related programmes run by other agencies, for example, agriculture, community development, religious bodies, various forms of private medicine (doctors in private practice, retail chemists, services provided by employers); hospitals which provide Primary Health Care. The task of a local management team is to identify what is already taking place, support and develop it and knit it into whatever new activities and services may be planned.

The greatest challenge is how to make the people feel that the health system is 'theirs'; that they own it, take pride in it, and support it, because it meets their needs and fits in with local traditions.

(5) *Community involvement.* The greatest resource available is likely to be the community itself and it can be the foundation on which the whole health system is built. Over the past 30 years many countries have made large investments to build up the industrial and commercial base of economic development in the belief that this would lead to the building up of an educated élite and an urban middle class and that consequent economic benefits would 'trickle down' to all sections of the community and act as a spur to development amongst the mass of the rural poor. This theory is now being questioned. What this approach to development ignored was that the greatest potential asset of any country lies in its own people's resourcefulness and their will to work for the improvement of their living conditions. To work well people need to be healthy, and this means that health services have a crucial part to play in the over-all development programme.

Moreover, health programmes can themselves be a 'trigger' for further development, by mobilising the resources and capabilities of communities to fulfil their own aspirations. Guidance and assistance can be welcomed by rural communities to build their own health services if it is genuinely offered in practical ways which the local population feel are appropriate. The skills and understanding of how to work *with* communities in this way are fundamental to successful Primary Health Care. (After all, getting things done with people is what management is all about.) The understanding is rooted in the academic disciplines of psychology, sociology, communication, anthropology and other human sciences which are being given increased attention in the training of health workers, supplemented by field work to understand community interactions in a local population and its ways of living. But the attitude is basically a personal one in which health workers say 'We have certain medical skills, and know-how about how people can keep themselves healthy, but it is the community who best knows how this knowledge can be applied. Therefore we need to get to know the community, be trusted by the people in it, and work *with* local people to provide the services they need'. Furthermore, involving the communities in the identification of their needs, planning health programmes, implementing and evaluating them, raises their level of awareness of their health and other problems and commits them to doing something about their problems.

(6) The *District Health Team* (DHT) is a key element in the thinking behind this book. A District Medical Officer (DMO) does not work in isolation.

Primary Health Care is a comprehensive system of care, prevention, treatment, community development, management and organisation. Yet few doctors possess all the skills and knowledge required, nor have control over all the health workers and resources in the District.

At the District level there may be such people as:

(a) Medical assistants, and other auxiliary workers.

(b) A senior public health nurse with responsibility for midwives, public health nurses and community nurses.

(c) A senior environmental health officer responsible for environmental health with a staff of public health inspectors, sanitarians, assistants, etc.

(d) A senior nutrition worker, responsible for monitoring the nutrition extension work at the health institutions and in the community.

(e) A Health Education Officer.

(f) Administrators/Finance Officers/Supplies staff with responsibilities for Primary Health Care.

Together such people can work as a team with one or more persons designated as managers. However, it should be recognised that management of Primary Health Care is a different management task from that of a hospital (see table 1.11).

Together such people can work as a team with one District Medical Officer accepting a joint responsibility for Primary Health Care. In addition such people as Community Development Officers, Education officials, Agriculture and Veterinary officers, community leaders, may work closely with a District Health Team. But it is the small 'core' DHT which is the 'nerve centre' and powerhouse for overseeing the work of the District and carrying out the functions of planning, developing, maintaining and evaluating services; supporting, building and developing staff and systems of work; and maintaining links between the local situation and government, regional and other agencies. Whilst having a joint responsibility for the District as a whole, each team member retains responsibility for his/her own specialised activities. The 'team' approach provides an opportunity for better joint decisions; consideration of a wider range of ideas and suggestions; support from members of the team to one another; unity, so that key staff and 'managers' in the District speak with a common voice; common policies agreed by all; continuity, whereby the team continues even if one person is absent; and a recognised means of settling differences and problems between people who work together.

Working relationships may need to be clarified where there is a District Hospital in the same District. A management team may be involved in running the hospital, for example, a Medical Director, Matron and Hospital Secretary, and there may also be some common membership between the Hospital Management Team and the District

Table 1.11 **Management of primary health care is a different kind of management task from that of the hospital**

Hospital	Health Centre has both features	Primary Health Care
Complex systems		Relatively simple system
High technology		Low technology
High capital investment		Low capital investment
High technical skill in small numbers of trained staff		More of low technical skill – widely distributed
Changing patient population – relatively shallow personal relationships		Stable community population – deeper personal relationships
Professionals do things to patients (passive patient)		Health workers do things *with* mothers, families (active client involvement)
High degree of specialisation – specialist doctors, nurses, technicians		Health workers are multi-purpose and do many tasks – diagnosis, treatment, clerking, social work, nursing
Short-term visible results, individual patients get better		Long term results, not always visible – the health of the community improves over a period of time
Disease centred – responds to disease demands presented to it		'Health' centred – identifies total community health *needs* and acts on them.
Prestigious. Patients and staff want to come to it		Low-key, mundane – some people want to get away from it
Comparatively rich		Poor – financially and educationally
Calls for high order administrative procedural skills (tendency to strict, autocratic management, e.g., in operating theatre)		Calls for high order personal and social skills and common-sense and adaptability. Needs open, encouraging, participative management style
'Alien' system introduced from outside – an urban institution		Integrated and woven into the fabric of the community
Grows from external resources		Grows from internal resources
Highly structured hierarchical organisation		Individuals and small groups work on their own and often in isolation
Creates its own dynamism		Dynamism must be maintained

Health Team. Where this is so, the functions and scope of each team will need to be defined clearly.

There can be special problems in relating the work of Primary Health Care to that of a hospital. A hospital's main function is to provide secondary care (or tertiary care). The nature of primary and secondary care is essentially different, and table 1.11 sets out some of the major differences managers may need to bear in mind.

Where manpower and other constraints do not permit, Health Teams can be set up to manage District health services.

(7) *A multi-level structure.* Remarkably similar structures are being developed in a number of countries in East and West Africa, Asia, and elsewhere. A basic 3–4-level system operates as follows:

Level A – Village or small community, serving a population of about 500–1000. The community is involved through a village health committee and village health workers working in association with traditional birth attendants and local healers. Services provided are mainly in the fields of prevention, hygiene and sanitation, first aid, simple diagnosis and treatment, antenatal and postnatal care, child care and control of communicable diseases.

Level B_1 – A sub-centre or health post serving a population of 5000–10 000, staffed by two to three paid workers, for example, rural medical and nursing aide, midwifery aides, and so on. It provides diagnostic and outpatient service, holding beds for acutely ill patients, antenatal and under-fives' clinics and midwifery services. Some countries have only a midwifery centre at Level B, whilst in others Levels B_1 and B_2 are combined for administrative reasons.

Level B_2 – Health Centre serving a number of sub-centres (in practice, any number from 5 to 20) and a population of about 50 000–100 000, staffed by a small number of paid workers, for example, a medical assistant, public health nurse/midwife, environmental health worker/sanitarian. The health centre provides clinic, diagnostic and treatment services for patients referred by village health workers and sub-centres. It also supports village health workers to ensure that as much work as possible is done at the village level.

Level C – The District level serving a population of between 200 000–500 000. A District Health Team is responsible for planning, administration and support of Health Centres, health posts and village health workers throughout the District, and providing help to patients and problems referred to them from Health Centres. Services may also be provided in a District Hospital or through a District Health facility in which more advanced services and treatments are provided than are possible in Health Centres. District mobile teams may provide services throughout the District and support the work of Health Centres.

Beyond the District level there is the National level (Ministry of Health) and in large countries an intermediate Provincial or Regional level. Variations of the basic 'multi-level' approach depend on geography, population, resources available, and so on, but it is important to incorporate the main principles of the approach in any local arrangement (see table 1.12).

(8) *Clearly understood roles.* Those who work within a Primary Health Care system must have a clear idea of what they are doing and trying to

Table 1.12 **Principles in establishing a multi-level structure**

(a) The village and local community as the foundation for all health services.
(b) Services provided at as local a level as possible, by health workers trained to make the greatest impact on the most common health problems of their locality.
(c) Support from more distant levels (C and B) to local levels (B and A) in the form of training, encouragement, supplies, etc.
(d) More distant levels deal *only* with the relatively few cases and problems which cannot be dealt with at local levels.
(e) Concentration of technical expertise at District level but readily available and constantly supportive for use throughout the District.
(f) A co-ordinated multi-disciplinary and multi-activity approach. Thus at village (local community) level, the village health committee (or in some cases village development committee) and village health workers are concerned with prevention, care, environmental health and simple treatments; at sub-centre and Health Centre level each worker has his or her special function, but with a good deal of inter-changeability; and at District level, the District Health Team of specialist professionals work together to plan and support a co-ordinated approach to the health needs of the District as a whole.
(g) Hospitals where they exist, are incorporated into the health care system of the District as a whole rather than operating as separate institutions.

achieve. A number of quite distinct roles can be identified; for example:
(a) Village health worker
(b) Traditional birth attendant
(c) Health Centre Superintendent who may be a doctor as in India, an auxiliary as in Tanzania, or a trained nurse as in Ghana.
(d) District Medical Officer
(e) District Co-ordinator of Mother and Child Health (MCH)

These roles are quite specific to Primary Health Care and need to be related to the needs and requirements of the particular local situation. Individuals holding these roles may have had formal training (for example, as a nurse or a doctor) but that training may not be sufficient in itself and will need supplementing and developing. Over a period of time conditions also change, and roles need to be flexible enough to adapt. Outlines are given of the main functions of a Primary Health Worker (see table 1.13) and of the managerial role of doctors in community medicine in developing countries (see table 1.15). Such roles would need to be understood by the individuals who hold them and by their supervisors, but also by other professionals, colleagues and the community with whom they relate so as to avoid unrealistic expectations.

(9) *Clearly defined systems.* Within this structure there are a number of *systems* which need to function well if good health care is to be provided. To use the analogy of the body, the 4-level structure provides the skeleton or anatomy of the District, but a number of systems also need

Table 1.13 **The primary health worker (PHW) profile**

PHW Organisational Relationships
The PHW is responsible both to the local community and to a supervisor appointed by the national health services.
The PHW follows the instructions given by a supervisor and will work with him or her as a member of a team.

Duties of the PHW:
1. Cares for the health of the members of his community and promotes community hygiene.
2. Gives care and advice during illness and arranges for referral if necessary.
3. Refers patients to the nearest Health Centre or hospital if they cannot be treated by him locally. The PHW should therefore confine care and treatment to those cases, conditions and situations for which he is adequately trained.
4. Where necessary, visits homes and gives advice on how to prevent disease and develop good habits of hygiene.
5. Makes regular reports to the local authorities on the health of the people and on the conditions of hygiene in the community. Gets the local authorities and the people to give him the help and support he needs for his work.
6. Keeps in regular contact with his supervisor so as to be able to give of his best in his work and to obtain the equipment and supplies he needs.
7. Promotes community development activities and plays an active part in them. This assumes that the PHW:
    (a) is available to respond to any emergency calls
    (b) acts in all circumstances with commonsense and in awareness of his or her limitations and of his or her responsibilities
    (c) does not leave the community without first informing the local authorities (village development and/or health committee)
    (d) takes part in the periods of training organised by the health service
    (e) motivates parents and families to make use of available health services e.g. MCH clinics.

The PHW may spend some time with other social/developmental workers involved in improving agricultural practices, storage of food, water supply, home economics, etc. The PHW needs to know about development opportunities in the district and must keep the community properly informed.

(Adapted from 'The Primary Health Worker Working Guide' pp. 3–5, WHO 1977).

to work effectively and in harmony with one another. Such systems include:

(a) Systems for diagnosis, referral, treatment and care of patients.
(b) Systems for antenatal, postnatal and child care.
(c) Systems for identifying and tackling the community's major health problems, for example, communicable diseases.
(d) Transport and communication systems.
(e) Systems to do with management of staff, for example, recruitment and training.
(f) Systems for the procurement and distribution of drugs, equipment and other supplies.

HOW HAVE COUNTRIES RESPONDED TO THE CONCEPT OF PRIMARY HEALTH CARE?

Since the Alma-Ata declaration on Primary Health Care, nations have been introducing a number of changes in their health policies and systems to reach the target of Health For All by the year 2000. Countries vary in the progress achieved because of differences in availability of resources and the state of development of the health systems. However, the following trends can be identified:

(1) In some countries PHC is the basis of the entire national strategy for health. In others it forms one component of the health policy.
(2) Support mechanisms for PHC at the National, Provincial and District level are being created. Such mechanisms include information systems, management of personnel and finance, procurement of essential drugs and medicinal supplies, maintenance of equipment, transportation including road systems, and communication.
(3) In the growing trend towards decentralisation under PHC the Districts are becoming the new focus for health development. New roles are being defined for the various hierarchical levels of the health services. The community (Level A) is the operational level; the health centre and sub-centre (Level B) have the support role for Level A; the District (Level C) has the micro planning and managerial role with the province (Level D) taking on the broader technical, administrative, training and other supportive functions. Macro planning, policy formulations, general administration and the allocation of resources according to need remain the prime responsibility of the Ministry of Health.

These new roles are being exercised with the involvement of health management committees at all levels of the health services. The activities of these committees go beyond the narrow definition of health, depending upon the level of sophistication of the communities concerned. For example, topics related to collaboration between sectors, food security, and the environment are also included in the activities of the committees.
(4) The majority of countries are using community-based health workers as part of their PHC strategy. Extensive experience has been gained regarding the training of community health workers, in producing curricula and training materials. It is now generally recognised that nurses, midwives and auxiliaries will continue to provide the greater part of PHC. Their roles are being further defined, and their training accordingly modified.
(5) Assessing the health status of individuals and communities, methods of community mobilisation, provision of integrated health care

including treatment of emergencies and referrals, maintaining epidemiological surveillance, collaborating with other development sectors, and maintaining progress in PHC are increasingly recognised as important activities. The training of members of the District Health Team in these topics is being given increasing prominence. The Districts are thus being empowered to become more resourceful in responding to disease outbreaks.

At the same time Health Service Research to monitor and evaluate the effectiveness and efficiency of the services being provided is being given priority. These new sets of priorities have identified the need for training in research planning and epidemiology.

(6) The need to upgrade and strengthen health infrastructure at the peripheral level is widely recognised. In a harsh economic climate of recession and punitive structural adjustment programmes this is still being accomplished, often through transfer of resources from the curative to the preventive/promotive services.

(7) Community involvement is being given emphasis through utilisation of existing local level infrastructure and personnel, for example Farmers' Associations; Women and Youth organisations; urban neighbourhood groups; traditional health workers; and so on. At the same time new categories of local level infrastructure and manpower are being created, for example village and neighbourhood development committees; local level insurance schemes, community health workers, and so on.

A number of new approaches in sustainable development through the community becoming a partner with the centre to share in capital expenditure and maintenance of the local infrastructure have been evolved. This also includes financial undertakings in the form of revolving funds for the procurement of medicinal supplies.

(8) Innovative approaches are being developed for food security at the level of the family, the community and the nation. Activities in connection with improving food supply and nutrition are being integrated into PHC in the form of nutritional surveillance and control of deficiency disorders. Intersectoral collaboration has come to play a significant role through linking of activities like fish farming, animal husbandry, demonstration farms, cottage industries, co-operatives for sharing of equipment and fertilisers as well as for marketing, and credit facilities.

(9) Programmes of immunisation (EPI), and for the promotion of Oral Rehydration Therapy (ORT) have made major impacts on child mortality and morbidity. Similar programmes for the control and treatment of acute respiratory infections and for safe motherhood are gathering momentum. These programmes represent the cutting edge of PHC.

(10) It is being increasingly recognised that concepts like PHC and Health For All will remain elusive as long as the health system is fragmented. Various sectors, institutions, and agencies are involved in the health development process. The vital connection between food supply, adequate nutrition and agriculture as well as employment; between maternal education and child health; between improved communication systems and peripheral health workers has to be recognised and nurtured through relevant policies.

DOCTORS AS MANAGERS AND LEADERS

Since the District has come to be increasingly recognised as the hub of the national strategy of Primary Health Care, new roles are being assigned to the District Medical Officer (or the equivalent) and the District Health Team for the implementation of this strategy. The most significant role of the District Medical Officer is that of being the manager and leader of the District Health Team.

One may be appointed a manager, but one is not a leader until the appointment is 'ratified' in the hearts and minds of the team one leads. What a leader should be good at is inspiring others. This is very much tied in with the leader's own enthusiasm and commitment, and with the ability to communicate and share that enthusiasm with others.

Leadership is also about teamwork, and creating teams. It goes without saying that leaders tend to create teams, and teams tend to have leaders.

Teams and working groups share three areas of common needs:

(1) The need to accomplish a common task
(2) The need to be maintained as a cohesive social unit
(3) The sum of the individual needs of those comprising the group

Failure in any one area affects the other two, for example failure to achieve the task will both disrupt the sense of teamship as well as lower the level of individual satisfaction.

Functions of leadership

In general the leader/manager of the District Health Team will be expected to carry out the following functions:

(1) *Planning*. Seeking all available information; defining tasks and goals for various teams; making a workable plan.
(2) *Initiating*. Briefing the team(s); allocating tasks; setting group standards.

(3) *Controlling*. Maintaining standards of work for the various teams; ensuring progress towards objectives; 'prodding' for actions and decisions.
(4) *Supporting*. Expressing appreciation of individual contributions; encouraging and disciplining; creating team spirit; relieving tension with good humour; reconciling disagreements.
(5) *Informing*. Clarifying tasks and plans; keeping the team(s) informed; receiving information from the various groups; summarising ideas and suggestions.
(6) *Evaluating*. Checking feasibility of ideas; testing consequences; evaluating team performances; helping team(s) to evaluate themselves.

With regard to team work it is always helpful to bear in mind the so called 50:50 Rule. The rule states that 50 per cent success depends on the team and 50 per cent on the leader. It has the advantage of challenging each party to get their own performance right before criticising the quality of contribution from the others.

The 50:50 Rule also applies to motivation. Half an individual's motivation comes from within himself, and the other half from external sources, including the leadership.

Decision-making

The manager/leader of the District Health Team will be called upon to make three main types of decisions viz. strategic, administrative and operating.

(1) *Strategic decisions* relate to the goals and objectives. They relate to new opportunities to be exploited as they arise, and new resources to be explored.
(2) *Administrative decisions* are about defects and deficiencies in the organisational structure, and how these can be improved to achieve the stated objectives.
(3) *Operating decisions* are about performance standards; about appraisal, and about monitoring and control.

Key abilities of a good leader

A number of abilities and skills are necessary for effective leadership. These are:

(1) *The management of attention*. The leader defines a vision of the future that the others can believe in and adopt as their own. With a vision the leader provides a bridge from the present to the future of the organisation (see table 1.14).

Table 1.14 A shared vision

A shared vision is a view of the future to which the District Health Team has made a commitment.

A shared vision is important because:
1. People then know what the organisation is trying to do and see how their own role fits into the general purpose.
2. There is a unity of purpose which reduces role and job misunderstanding.
3. A clear direction helps to challenge the creativity of the workers.
4. The energies of the work force are channelled in the direction of achieving a clear aim.

Vision is derived from:
1. An analysis of the past taking account of the working pattern and traditions of the organisation.
2. An analysis of the present situation so that key factors influencing the successes and failures of the organisation are understood.
3. An analysis of the future needs and looking at all aspects that could influence the organisation.

Vision statements should be:
1. Clearly expressed in simple language and avoiding ambiguities.
2. A reflection of what is required.
3. Intellectually stimulating and exciting.

---

(2) *The management of meaning.* Communicating the vision and translating it into successful results in the District health activities.

(3) *The management of trust.* Leaders must be consistent in order to generate trust, which acts as glue that binds followers and leaders together.

(4) *The management of self.* Persistence, self-knowledge, willingness to take risks, commitment and challenge. The most important quality is the willingness to go on learning, particularly from setbacks and failures.

Taking the above description into account one can now list the functions of a manager of the District Health Team. These can be listed as follows;

(1) *A manager sets objectives* in line with the vision of the future. This helps to determine what the goals in each area of the objectives should be and to decide what needs to be done to reach these objectives. The manager makes the objectives effective by communicating them to the people whose performance is needed to attain them.

(2) *A manager organises.* This requires classifying the work to be done and dividing it into manageable activities, as well as further dividing the activities into manageable jobs. The manager then groups these units and jobs into an organisation structure and selects people for the management of these units and for the jobs to be done.

(3) *A manager motivates and communicates.* The challenge is to make teams out of people that are responsible for various jobs. This is achieved through constant communication to and from the team members, and to and from the superiors as well as from one's colleagues.
(4) *A manager provides for measurement.* Quality of work can be assured by establishing yardsticks, and ensuring that each member of the workforce has these standards available to them. It is the function of the manager to analyse, interpret and appraise performance. As in all areas of work the manager communicates the findings to the team members, to superiors, and to colleagues.
(5) *A manager develops people* through organising training in the form of regular study days, workshops, secondments and so on.

What are the desirable traits of a successful leader?

Most studies have been carried out in the industrial and commercial sectors. These single out the following qualities:

(1) Above average intelligence, particularly the ability to solve complex and abstract problems.
(2) Initiative, independent thinking and inventiveness.
(3) Qualities of imagination and vision tinged with pragmatism enabling the person to not only sum up the situation but also attainable future goals.
(4) Initiative. The capacity to perceive a need for action and the urge to do it.
(5) Self assurance with high self ratings on competence and aspiration.
(6) Ability to develop consensus rather than wanting to control things, or dealing with mistakes.
(7) Other abilities described are enthusiasm, sociability, courage, imagination, determination, and energy.

Factors influencing the work of the district manager

The main forces influencing the work of the District Manager are threefold:

(1) *Policy guidelines from the Ministry of Health.* These are influenced in turn by international trends, recommendations of international agencies like the World Health Organisation, the United Nations Children's Fund, local academic and research institutions, as well as available funds.

   The District Health Team would be able to influence the policy guidelines if they have set up a good information system within the

District and the pattern of locally endemic diseases has been carefully researched.
(2) *The health strategy for the region (or the province) and of the District.* The strategy has been worked out along the lines of the national policy guidelines to establish long term goals and objectives, courses of action, and allocation of resources. The District Health Team has ample scope to contribute to the health strategy of their District.
(3) *The organisational structure within the District.* The organisational structure is formulated to translate the strategy into service delivery. It is largely the responsibility of the District Health Manager to evolve an effective organisation. Without the right organisation great ideas will only breed great frustration.

The conceptual model we all carry in our minds about an organisation is that of a machine with interlocking parts. In this model people are looked upon as 'human resources'. There are 'plans', 'control systems' and 'inputs' and 'outputs'. But the language of management is changing rapidly, and there are other ways of looking at organisations. They can be conceived as communities with a common purpose, made up of individuals with minds and values of their own. In such communities there are 'shared values', 'networks and alliances', 'compromise and consent', and leadership more than management.

Management styles

Management styles are very personal, and vary from one manager to another. Any one of the following or a mixture of one or more management style may be found in a given District:

(1) *Exploitative authoritarian.* Communication is top down and decision-making is done at the top with no sharing of ideas with those in the front line. In such a situation the superiors and subordinates are psychologically far apart, and management is by fear and coercion. (The manager prefers to make the decision or solve the problem on his own using the information available at the time.)
(2) *Benevolent authoritarian.* Policy decisions are still being made at the top, and only minor ones are delegated to a lower level. Management is by carrot rather than stick, and such information as flows upwards is mainly what the bosses are thought to wish to hear. (The manager obtains the necessary information from the subordinates, then decides on the solution to the problem himself.)
(3) *Consultative.* Management does try and talk to employees. Communication flows both ways but is still somewhat limited upwards. Important decisions are taken top down, and management uses both carrot and stick. (The manager shares the problem with relevant subordinates

Table 1.15 Analysis of the managerial role of doctors in community medicine in developing countries

Content of work

1	Technical skills	e.g. Diagnosis and Treatment of: (a) Individuals (b) Communities
2	Teaching	i.e. Giving other people the knowledge and skills to perform competently
3	Management of resources	Management of finance, e.g., general budgeting and financial control. Personnel management, e.g., arrangement of clinic sessions Disposition of Vehicles Allocating and delegating work to others
4	Data gathering	e.g. Demographic profile of the district; epidemiological characteristics and health statistics.
5	Planning	Thinking and planning ahead to determine what should be done in the future.
6	Innovation and development	i.e. Introducing new approaches and development to achieve better results e.g. Encouraging, supporting the work of others. Giving a sense of direction.
7	Community mobilisation	e.g. Involving the community on the day-to-day level in planning implementation and evaluation of health programmes.
8	Minor administrative matters	
9	Dealing with crises	

individually or in a group, obtaining ideas and suggestions. Then the manager makes the decision.)

(4) *Participative*. Management works closely with employees to encourage high performance. Communication flows easily in both directions and sideways to peers. Managers and workers are psychologically close. Decision-making is by a participative process. Working teams are integrated in the management structures by means of team leaders being both a member of the team and management. (The manager shares the problem with the subordinates as a group. Then together they come to a decision.) (See also page 27.)

In real life decision-making is rarely stereotyped, and will depend upon a number of factors: for example, the *leader* and his preferred style of operating and personal characteristics; the *task in hand*, its objective and the available technology; and the *environment*, i.e. the organisation in which the leader and the team are performing the task. That being the case, there cannot be such a thing as the 'correct' style of leadership. Leadership is most effective when the requirements of the leader, the team, and the task fit together in the environmental setting which includes the norms of the organisation, the technology and the variety of tasks.

FURTHER READING

Ebrahim G. J. A model of integrated community health care in a rural area. *Trop. Geog. Med.* (1976) **28**:Supplement.

Ebrahim G. J., Kiango A., Twaha A. Learning from doing: progression to Primary Health Care within a national health programme. *J.Trop.Ped.* (1988) **34**:4–11.

Twaha A., Ebrahim G. J., Vogel G. C. J., van Prag E. Lessons for national health systems from small scale projects: a case study from Tanzania. *J.Trop.Ped.* (1989) **35**:40–43.

World Health Organization. *Primary Health Care*. WHO 'Health for All' Series No. 1, WHO, Geneva, 1978.

World Health Organization. *Managerial Process for National Health Development*. 'Health for All' Series No. 5, WHO, Geneva, 1981.

2 Finding Out About Health Needs in the District

> When therefore a physician comes to a District previously unknown, its situation and its aspect to the winds must be considered. This is of the greatest importance and the effect of each season of the year must be studied. Similarly the nature of the water supply must be considered; then the soil, whether it be bare and waterless or thickly covered with vegetation. Is the area hollow and stifling or exposed and cold. Lastly consider the life of the inhabitants themselves; are they heavy drinkers and eaters and consequently unable to withstand fatigue or, being fond of work and exercise eat wisely and drink sparely?
>
> Hippocrates: *Airs, Waters, Places*

The five key questions in community diagnosis of sickness are raised in this quotation. Who becomes sick? Where? When? With what? and Why? The same five questions are also applied to identifying health care resources; Who is providing care? What? Where? When? And we need to ask Why? as well for there are many different reasons why a particular pattern of health service develops. In order to decide what actions are needed in a District these two sets of five questions need to be asked to identify the specific problems and the resources available to deal with them.

In addition, community diagnosis needs also to describe community factors. It needs to describe the pattern of disease not just in terms of types of disease but also in terms of causation, for example, infectious agents and their prevalence, food production, life-style, beliefs and attitudes, and so on. It is not enough simply to say that people do not cultivate enough food. Answers are also needed to questions such as 'Why aren't people cultivating enough food crops?' 'Is the soil unsuitable?' 'Is there a shortage of land for some people?' 'Is there more scope for change within the community in relation to some things, for example, "soft points" compared with other aspects which are far less likely to change, that is, "hard points"?'

WHO? WHAT? WHERE? WHEN? WHY? IN ILL HEALTH

Which age groups contain most people and which age group is increasing fastest?

The age structure of the population in a District is the first indicator of what the pattern of health problems is likely to be. In many developing countries children under five make up about 20 per cent of the population and women and children together often account for as many as 65 per cent of the people. Older people, over 65, are also a rapidly increasing age group all over the world. The precise situation in a District can be obtained from census data corrected for the annual growth rate since the census was taken. In one District in Ghana, Ashanti-Akim, it was found that women of child-bearing age and children under 15 made up 65 per cent of the population (see table 2.1). Such a pattern of population is typical of most developing countries and is the result of a high birth rate coupled with low life expectancy.

Table 2.1 **Population of the district from census data (1970) (corrected, allowing for 3% growth rate per annum) Ashanti-Akim District Profile, 1979**

Population of district	156 000	
Total under 5	29 299	(19%)
Total under age 15	78 000	(50%)
Women of child-bearing age	23 217	(15%)
Total Women of child-bearing age and children under 15	101 217	(65%)

Who gets sick? Who dies?

The heavy burden of ill health borne by children can often be seen from the high proportion of child deaths amongst the top ten causes of deaths occurring in a District Hospital. Nearly a quarter of hospital deaths may be children. For every child that dies there is a bereaved family in need of emotional and social support. Although maternal deaths may thankfully be less frequent, when a mother dies several children may suffer and a household may lose a major contributor to its labour force.

Tables 2.2 and 2.3 give the most common causes of death in hospitals in Ghana and in Tanzania. In the case of the former, up to one in four deaths were in children and in the case of the latter, one in three deaths occurred in children. Disease problems often occur more frequently in certain age-groups, and in certain families, as well as in those who have been frequently ill before. Malnourished children are far more likely to fall ill, develop serious illness and

Table 2.2 **Top ten causes of death in one year – Agogo District Hospital, Ashanti-Akim District Profile, 1979 (N = 5056 admissions)**

	% of total
Premature and neonatal	13.9
Kwashiorkor malnutrition	10.0
Heart disease	8.9
Liver disease	7.9
Pulmonary disease	6.7
Malaria	6.3
Septicaemia	6.2
Intestinal disease	6.1
Tuberculosis	5.4
Measles	4.6

Table 2.3 **Most common causes of deaths in hospitals, Tanzania 1972**

	% of Total
Pneumonia	15.6
Diarrhoeal disease	9.6
Malaria	4.4
Tuberculosis	4.7
Diseases of the heart	4.5
Meningitis	0.9
Defective nutrition	7.1
Anaemia	4.8
Conditions of early infancy	6.9
Measles	10.5
Tetanus	4.6

succumb to it compared to children who are well nourished. Malnourished children under the age of five are particularly prone to diarrhoea. In a recent survey of Kwale District in Kenya (population 267 000) diarrhoea was the main complaint for which children were brought to its 121 health units and occurred in 15 per cent of 235 children under the age of five years. Diarrhoea did not feature at all as a major complaint in 148 children aged 5 to 15, nor in 200 adults.

Who needs maternity care?

The size of the need for maternity care in the District is partly indicated by the number of births. This figure can be obtained by using the National crude birth rate and the District population to calculate the District births expected. This can be compared with the District total of registered births. Often there are discrepancies between these numbers since in many countries few births are registered.

Table 2.4 **Number of births expected and found registered (1978) Ashanti-Akim District 1979**

Expected births	Registered births in the District	Per cent of District births registered
National crude birth rate = 50 per 1000 population		
District population = 156 000		
Expected births in the District = 7 800	2 109	27%

What can be done about high maternal mortality?

Over her lifetime the averge mother in the developing world (excluding China) faces one chance in 33 that pregnancy or childbirth will cause her death. By contrast women in the more developed countries face only one chance in 1500 of dying in pregnancy or childbirth. Recent estimates are for over 600 mothers' deaths per 100 000 live births for Africa, over 400 for Asia, and nearly 300 for Latin America and the Caribbean. On aggregate there are half a million maternal deaths each year. Of these 99 per cent occur in developing countries. These tragic statistics go even further. Each woman who dies in childbearing age leaves behind on average two or more motherless children. For every woman who dies in childbirth there are several more who are incapacitated or disabled. In both instances the survival chances of the children are seriously affected, since the health of young children is heavily dependent on the continued good health of their mothers.

Three quarters of all maternal deaths are due to five complications arising during pregnancy or labour viz. haemorrhage, obstructed labour, infection,

---

What is a maternal death?

Maternal death is death of a woman while pregnant, or within 42 days of the end of pregnancy (including termination), irrespective of duration or site (e.g. ectopic) from any cause related to or aggravated by the pregnancy or its management.

toxaemia (hypertensive disease of pregnancy), and septic abortion. Of these haemorrhage, infection, and hypertensive disease of pregnancy together make up half of all maternal deaths. Most of the complications are seen in the 'unbooked' cases, in whom the risks are more than 50 times those in the 'booked' healthy mothers. This indicates that besides the 'direct' causes listed above there must be some 'indirect' causes like failures and weak links in the health system, poor transport and communications facilities, socio-cultural blocks as well as political obstacles.

Table 2.5 **Causes of maternal mortality**

Direct causes	*Indirect causes*
Obstructed labour	Lack of antenatal care
Ruptured uterus	or poor quality of care
Antepartum haemorrhage	Poor roads and transport
Placenta praevia	Social and cultural difficulties
Postpartum haemorrhage	Planning and political neglect
Retained placenta	
Abortion	
Hypertensive	
disorders of pregnancy	
Peurperal or	
post-abortion sepsis	
Ectopic pregnancy	
Severe anaemia	

Facilities needed at the first referral unit

Surgical and obstetrical skills
Monitoring in labour and other midwifery skills
Anaesthetic skills
Medical skills
Management of high risk women
Blood replacement facilities
Family planning
Postnatal care
Neonatal care

Common clinical presentations

Pain	Infection
Dehydration	Convulsions
Blood loss	Full bladder

Life saving interventions

Intravenous fluids and drip sets
Plasma expanders
Blood transfusion facilities
Drugs (antibiotics/pethidine/diazepam)
Catheters

Requirements for safer motherhood

Like all programmes for health development safer motherhood must also begin with creating awareness amongst the public, health planners and the professionals. The aim is to recruit popular and professional support for creating a tiered system of maternity care which has its roots in the community with trained birth attendants and community midwives working out of health centres, health posts, and outreach clinics. A backup referral system based at Health Centres and District Hospitals for receiving high risk pregnancies and emergencies, as well as a reliable communications system including transport for high risk and complicated deliveries is equally important.

The strategy in a multi-tier system of care is to extend antenatal coverage to the maximum in order to identify early the quarter or so pregnant women in whom 75 per cent of life threatening complications occur. If such a system can also respond to the unmet needs of family planning then additional benefits from reduction in deaths due to unskilled abortions can be expected.

In such a tiered system of care the strategic decisions to be made by the District Health Managers are as follows:

(1) What interventions can be maximally effective at what levels?
(2) How can the system of care be further strengthened by means of technical back-up, regular training, and effective communications up and down the system?
(3) How can continuing improvements be brought about through regular audit and periodic evaluation?

In a systems approach the weakest element determines the overall strength of the system. No programme can work effectively if inputs and activities are concentrated at one level. A systems approach raises the question of tasks and functions at different levels; the training and skills of workers at all levels appropriate to the task they are expected to perform; the organisation of relevant training of staff members; the development of protocols for dealing with complications and emergencies at all levels; and adequate equipment to deal with life threatening situations (see table 2.6 and figure 2.1).

What are the health problems?

It is the serious diseases (as well as accidents which are common and either treatable or preventable) that are most likely to yield best results in response to intervention programmes. In children, nearly 40 per cent of the mortality is due to the major three diseases – malnutrition, respiratory infection and diarrhoea. About two-thirds of the mortality is due to the dominant nine which include the above three together with anaemia, tuberculosis, malaria and other parasitic diseases, whooping cough, measles and other common infectious illnesses of childhood, as well as accidents and poisoning.

Table 2.6 **Essential elements of obstetric care: a checklist**

Intervention	Available at Level A	Level B	Average distance from Level A to Level B

Surgical treatment
Caesarean section
Surgical treatment
  of sepsis
Repair of high vaginal
  or cervical tears
Laparotomy for repair
  of uterine rupture
Hysterectomy
Laparotomy for
  ectopic pregnancy
Evacuation of uterus
  for abortion
I.V. oxytocics to
  augment labour

Anaesthesia
General
Local
Resuscitation

Medical treatment
For sepsis
For shock
For hypertensive
  disease of pregnancy
Severe anaemia
Blood replacement
Fluid replacement

*Manual procedures
and monitoring of labour*
Manual removal
  of placenta
Oxytocics to prevent PPH
Exploration of uterus
Vacuum extraction
Partograph

Monitoring pregnancy

Maternity hostels

Special care of the newborn
Low birth weight
  and preterm babies
Birth trauma
Sepsis
Resuscitation
Special care

Figure 2.1 Essential elements of obstetric care related to the major causes of maternal mortality*

Essential elements of obstetric care at first referral level	Major causes of maternal mortality							
	Obstructed labour	Ruptured uterus	Antepartum haemorrhage	Postpartum haemorrhage and retained placenta	Abortion	Hypertensive disorders of pregnancy and eclampsia	Puerperal or post-abortion sepsis	Ectopic pregnancy	Severe anaemia
Surgical obstetrics									
Caesarean section	•	•	•						
Surgical treatment of sepsis	•	•					•	•	
Repair of high vaginal and cervical tears				•					
Laparotomy for repair of uterine rupture/ hysterectomy		•					•		
Removal of ectopic pregnancy presenting as 'acute abdomen'								•	
Evacuation of uterus in abortion					•		•		
Intravenous oxytocin infusion to augment labour			•	•	•				
Amniotomy with/ without oxytocin infusion						•			
Anaesthesia									
General						•			
Local	•	•			•	•		•	
Resuscitation									•

Medical treatment
For sepsis
For shock
Hypertensive disorders of pregnancy including eclampsia
For severe anaemia

Blood replacement
Fluid replacement

Manual procedures and monitoring labour
Manual removal of placenta
Oxytocics to prevent PPH
Exploration of uterus
Vacuum extraction
Partograph

Monitoring pregnancy
Waiting homes Places where women at high risk of complications can stay and receive supervision during the last month of pregnancy

Family planning support
Tubal ligation, vasectomy
Intrauterine contraceptive device (IUD)
Oral, injectable and implantable contraceptives
} To prevent unplanned or unwanted pregnancies which result in any of the above reasons for maternal mortality

Special care of the newborn
Premature and low birthweight babies, and those who have suffered trauma during delivery, will need special care including resuscitation and thermal control.

* A filled circle indicates that the element of obstetric care specified in the left-hand column is essential for the prevention of maternal mortality due to the named obstetric complication.

Source: Adapted from *Essential elements of obstetric care at first referral level*, WHO, 1991.

In mothers, health problems surrounding childbirth are likely to be of major importance, particularly nutritional deficiency, anaemia, malaria, puerperal sepsis including tetanus and obstetric accidents. In adult males, communicable diseases, particularly tuberculosis, nutritional deficiencies and accidents may be important.

However, in no country's statistics is it ever mentioned that the main health problem is 'Lack of Services', especially in rural areas and for the urban poor.

The key questions for the District manager are 'How can one find out which of these problems are important in a particular District' and 'How can appropriate services be developed to deal with them?'

Finding out about health needs

Vital statistics and morbidity data have been used traditionally to assess health needs. Thus, rates of infant and pre-school mortality, perinatal and maternal mortality and rates of deaths from specific diseases (for example, tuberculosis) have been used as a measure of the health status of the population. Unfortunately, in most developing countries exact vital statistics at the District level are non-existent. Births and deaths may not be recorded, and many illnesses are never brought to the attention of the health personnel. In such cases one can make a start by making estimates from national data if available, or go by any information or indicators reported in the literature on defined communities in the country. Alternatively, it may be possible to carry out a small scale study in one or more randomly selected villages in the District in order to get some impression of probable rates.

In one community in Madang Province in Papua New Guinea, some nursing sisters at the Catholic Mission on Manam Island obtained as much information as they could on all the deaths on the island which came to their attention in $5\frac{1}{2}$ years (July 1971 to December 1977, omitting 1974). The total deaths recorded indicates marked under-reporting but even then their data represent a much higher proportion of the expected deaths than is normally found from records of health units. Table 2.7 shows the data they collected. Childbirth accounted for the death of 4 women out of the 151 adults recorded, 3 per cent of adult deaths. Unfortunately we do not know the number of births in this $5\frac{1}{2}$-year period so we cannot calculate the maternal mortality rate exactly, but the figures show that maternal deaths are a very important problem in this area as in many others. Deaths are also important because where there is one death there may be several other people severely ill. Problems following childbirth such as severe anaemia, vesico-vaginal fistulae, infection and perhaps infertility, are likely to be more common when there are deaths in labour. The opposite is also true.

The various sources of data available in the District need to be used carefully if we are to find out the true picture of illness in the community. Certain problems will feature more in hospital statistics, others more in

Table 2.7 **Causes of death on Manam Island, Madang Province 1971–1977** (omitting 1974)

Adults (N = 151)	%	Children (N = 88)	%
Resp. Dis. (incl. Tuberculosis)	34	Malaria	27
Old age	13	Resp. Dis. (incl. Tuberculosis)	23
Cancer	12	Diarrhoea	15
Accidents	7	Heart disease	3.5
Heart disease	5	Epilepsy	3.5
Childbirth	3	Accidents	2
Other	24	Other	26
Diarrhoea	2		

Source: Madang Province Health Team Report, 1979.

Table 2.8 **Causes of death in Madang Hospital recorded on death certificates (January–December 1977)**

Adults (N = 118)	%	Children (N = 108)	%
Pneumonia	21	Septicaemia	12
Chronic Lung Disease	11	Pneumonia	10
Tuberculosis	8	Meningitis	10
Trauma	8	Trauma	5
Heart Disease	7	Gastroenteritis	4
Renal Failure	5	Malaria	3
Cancer	5	Malnutrition	3
Liver Disease	4	Perinatal	44
Other	3	Other	9

Health Centre data and some important problems we will only discover by visiting people at home.

In Madang District registered deaths among adults and children at the District Hospital are reported separately as in table 2.8. It was found that a very high proportion (44 per cent) of child deaths occurred in the perinatal period.

Health centre deaths reveal the seriousness of chest infections

In Madang District Hospital mortality data were supplemented by information on deaths at Health Centres (see table 2.9). Pneumonia was clearly a far greater problem in the community than had been indicated from the hospital

Table 2.9 **Causes of in-patient death in health centres, Madang Province (January 1975–February 1978)**

Adults N = 161	%	Children N = 139	%
Pneumonia	21	Pneumonia	44
Tuberculosis	16	Meningitis	9
Cancer	10	Gastroenteritis	8
Liver Disease	9	Meningitis/Cerebral	
Chronic Lung Disease	7	Malaria	8
Gastroenteritis	5	Malaria	7
Meningitis/Cerebral		Anaemia	5
Malaria	4	Trauma	4
Meningitis	3	Liver disease	4
Asthma	4	Tuberculosis	2
Childbirth	3	Malnutrition	2
Other	18	Other	7
		Malnutrition was associated with another 18 deaths (13%)	

Source: Death certificates returned to the provincial office.

data. It now accounted for 44 per cent of child deaths at the Health Centre, compared with 10 per cent of child deaths at the hospital. There were also more deaths at the Health Centres from diarrhoea, and malnutrition was associated with 13 per cent of deaths in addition to the 2 per cent where it was recorded as the primary cause.

Health Centre attendance or discharge data reveal high prevalence of diarrhoea, skin problems and abdominal pain and a higher frequency of malaria, bronchitis and trauma

Data on deaths only give a partial picture of the diseases prevalent in the community as a whole. In the Ghanaian District it was only when attendance data were studied from the Health Centres that the widespread problems of diarrhoea, skin disease and abdominal pain became apparent besides malaria and upper respiratory tract infections identified in the hospital earlier (see tables 2.2 and 2.10).

In the Papua New Guinea Health Centre, discharge data also revealed the high prevalence of diarrhoea in children and showed malaria to be far more frequent than death certificates suggested (see table 2.11). In adults, the discharge data show the importance of malaria, bronchitis and trauma (in addition to the pneumonia and tuberculosis shown in the mortality data in table 2.9).

The above discussion shows that every recorded death and every clinic attendance represents the 'tip of the iceberg'. For every known death or

District health needs

Table 2.10 **Top nine reasons for attendance at an outpatient clinic at Juaso rural health centre, Ashanti-Akim District Profile, 1979**

		% Total attendances
1	Malaria	31
2	Upper respiratory tract infections	15
3	Diarrhoea	12
4	Skin disease	7
5	Abdominal pain	6
6	Eye problems	3
7	Measles	2
8	Worms	2
9	Wounds	0.5

Table 2.11 **Discharge diagnoses from health centres (April 1978) Madang Province, Papua New Guinea**

Adults (N = 174)	%	Children (N = 183)	%
Pneumonia	25	*Malaria	25
*Malaria	20	Diarrhoea	21
Trauma	10	Pneumonia	14
Bronchitis	8.5	Malnutrition	12
Anaemia	8.5	Measles	9
Abscess	8.5	Trauma	4
Sores and ulcers	5	**Upper Resp. Tract Inf.	4
Diarrhoea	5	Eye infection	3
**Upper Resp. Tract Inf.	3	Sores and ulcers	3
Cancer	2.5	Anaemia	2
Other	4	Other	3
	100%		100%

* Includes fever, fever/headache
** Includes cough, cough/fever
This table does not include 86 confinements.

attendance at a clinic due to a particular cause there are likely to be several individuals ill for the same reason in the community. Thus known deaths and attendances can only give a glimpse of the picture of disease in the community. Clinic statistics from hospitals and Health Centres are notorious in under-reporting certain conditions, particularly chronic sickness, non-acute and non-epidemic diseases. Sickness rates are higher in the community than are ever reported at health units but clinic data give us a first view of the pattern of ill health.

Table 2.12 **Where are health problems in the district? Outpatient diagnoses (expressed as % of patients seen in each geographical region) Madang Province, Papua New Guinea 1978**

	Coast (2657 patients)	Ramu valley (1054 patients)	Mountain (985 patients)
Upper Resp. Tract Inf.**	12.5 ⎫	15 ⎫	21 ⎫
Pneumonia	4 ⎬ 24	5.5 ⎬ 20.5	3.5 ⎬ 30.5
Influenza	7.5 ⎭	– ⎭	6 ⎭
Sores and ulcers	16.5	25	22
Malaria*	25	24.5	8
Scabies	4	9	7
Diarrhoea	5	6	7
Trauma	4	1.5	5.5
Eye infection	2	2	1.5
Abscess	2.5	1.5	2.5
Other	17	10	16

* Includes fever, fever/headache
** Includes cough, cough/fever

Table 2.13 **Where are the malnourished children? Survey of 5 areas in Ashanti-Akim District, Ghana, 1977**

Arm circumference	Agogo (N = 599)	Bompata (N = 237)	Akutuase (N = 101)	Ananekrom (N = 60)	Nyaboe (N = 157)
≥ 13.5 cm.	75%	74%	65%	73%	59%
12.5–13.5 cm.	21%	19%	25%	20%	28%
≤ 12.5 cm.	4%	7%	10%	7%	13%
Total percentage ≤ 12.5–13.5 cm	25%	26%	35%	27%	41%

Where are the health problems in the district?

Very often ill-health clusters in certain places or communities or families. All front-line health workers need to be trained to always watch for this clustering phenomenon. This is what one nurse noticed in the course of her work in a peripheral health unit in Malaysia (adapted from *The use of epidemiology by front-line workers in developing countries*, World Health Organization SHS/SPM/81.3. 1981).

She had noted from the weight-for-age charts that there were six young children with undernutrition in her area. These children all came from two villages which were situated on rubber estates in her area. The other two villages in her area had no undernourished children.

She investigated the families of the undernourished children and found that, in most cases, both father and mother were rubber tappers who left for work early in the mornings. Their young children were left in unsupervised nurseries where they were only fed milk from baby bottles if they cried. The milk was prepared by the mother before she left for work and was not properly refrigerated. The children frequently suffered from diarrhoea. From her observations she made efforts to visit the nurseries, train the attendants and improve their hygiene and methods of feeding the children.

This health worker had asked the right questions.

What is the event? A child persistently below the 80 per cent line of weight for age.

Who are affected? Children of mothers who work as rubber tappers.

Where are the events? Two villages in rubber estates; children attending unsupervised nurseries.

When did the events occur? Continuing at time of contact with the community nurse.

Are these events usual? Not usual in other parts of the area.

Why have these events occurred? Working mothers, bottle-fed children, inadequate nurseries, infection and diarrhoea.

What are the key factors? Lack of understanding of hygiene and nutrition by nursery attendants.

The clustering phenomenon is best revealed by charting events on a map of the area. Community nurses often have the knowledge required to chart health events on a map or on a chart, to see whether they are concentrated either in space or in time, or in a certain type of family. And if they are, she asks herself 'Why?' For a tentative answer to this question, she uses other bits of information which are available to her 'Could it be that communication difficulties, or low family income, or illiteracy, or the presence of a particular illegal injectionist are influencing the events?'

Many Districts contain a variety of geographical areas and breakdown of data by these areas may reveal important differences due either to geographical or socio-economic factors. Table 2.12 shows outpatient diagnoses recorded in a specimen week in April 1978, in Papua New Guinea. They were obtained by calling in daily rolls to the Provincial Health Team Centre from all Health Centres, sub-centres and aid posts. Records were obtained from all Health Centres and sub-centres and 60 per cent of functioning aid posts. Because a Health Centre's outpatient department performs a similar function for its surrounding population as does an aid post, the results have been combined. The returns are classified by coastal, valley or mountain areas. Upper respiratory tract infection was rather more common in the mountain area, sores and ulcers were high in both mountain and valley areas. Malaria was nearly always found on the coast or in the valley, probably on account of conditions being more favourable for the mosquito to breed. Trauma seemed slightly less frequent in the valley populations.

In the Ashanti-Akim District of Ghana far more malnourished children were found in one area (Nyaboe) than in the other four areas studied (see table 2.13).

Census data, routine statistics and survey data, if available, can be used increasingly to identify areas of a District where health problems are more likely to occur. From such sources one may be able to gather indicators of 'social malaise' which are highly associated with one another and with specific health problems. Useful data and indicators may include the following:

(1) Differences in female/male mortality rates in specified age groups.
(2) Proportion of females/males literate in rural areas.
(3) Attendance rates at specified levels in schools.
(4) Median years of education completed, male and female.
(5) Percentage of women heads of household (widowed, divorced, separated, or husband away for long periods).
(6) Average number of children in women-headed households.

When does ill-health occur?

Many diseases are seasonal. In countries with dry and rainy seasons, each season may bring a different disease pattern. Meningitis may be rampant in a savannah dry season as in the cerebro-spinal fever belt of Africa, and malaria when the rains arrive. Often long-awaited rain is associated first with relief because crops will begin to grow, but then with concern as malaria increases and the mosquitoes breed in the puddles. Diarrhoea increases, too, as the drainage system cannot cope with the heavy downpours. The situation is exacerbated by the fact that for many communities the first rains also herald the beginning of the busiest time of the year in the fields planting the new crop. There is little time left to care for children at home. It is also often a hungry season with the new crop only just being planted and stores from the previous crop fast running out. Where food stores from the previous harvest are low, prices of food in the market are usually at their highest point. Knowing when ill health occurs, the high risk times of the year can be recognised and some of the factors associated with ill health identified too. In the example described above, one most effective intervention may be to increase storage capacity of food after the previous harvest so the length of the hungry season is reduced and people are less vulnerable to infection.

In Bangladesh, studies in Matlab Thana have shown seasonal changes in the price of rice, the agricultural wage and household stocks of cereal. In the hungry season, as household stocks declined, wages decreased and rice prices increased (see figure 2.2).

Rapid epidemiological assessment

Health managers need sound epidemiological information for forward planning, for deciding priorities, and for best use of resources. Unfortunately, classical epidemiological techniques are too cumbersome, costly, and require

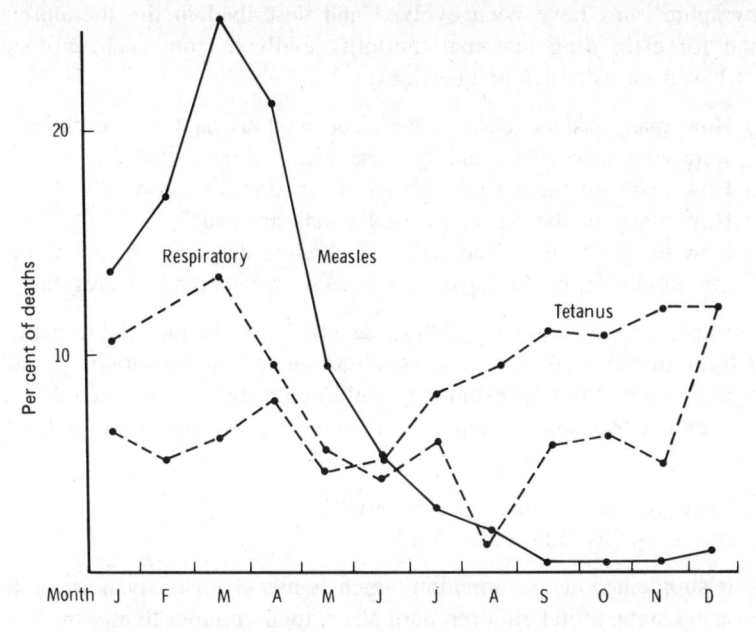

Per cent of child deaths (0-4 years) due to tetanus, measles, and respiratory diseases by calender month

Figure 2.2. Seasonal mortality in Matlab Thana area, Bangladesh, 1976

Source: Sowie, Chen, L. C. Rahman, M. Sarder, A. M. *Int. of Epidem.* (1980).

special skills. In recent years a number of techniques for rapid appraisal have been developed which are easy to apply. They provide near estimates and trends rather than exact figures. Even then they fulfil many of the requirements for quick decision-making.

The techniques for rapid epidemiological assessment fall into six general groupings:

(1) Prevalence rates of maternal and childhood mortality
(2) Prevalence rates of disability
(3) Sampling methods
(4) Methods of surveillance and monitoring
(5) Indicators of risk and health status
(6) Case-control methods

Details of each method can be found in the references at the end of the chapter. Only brief descriptions are provided in this section.

Prevalence of maternal and child mortality

Over the last 20 years a number of indirect techniques for estimating basic demographic rates have been evolved and described in the literature. The method for estimating maternal mortality evolved from such approaches, and is based on asking four questions:

(1) How many sisters (born to the same mother) have you ever had who were ever married (including those who are now dead)?
(2) How many of these ever married sisters are alive now?
(3) How many of these ever married sisters are dead?
(4) How many of these dead sisters died while they were pregnant, or during childbirth, or during the six weeks after the end of pregnancy?

A sample size of 2500 to 3000 is desirable to obtain reliable estimates. Field trials of the method in The Gambia have given encouraging results.

A simplified method for estimating child mortality has also been described. Two specific questions are put to a mother at or soon after a delivery as follows:

(1) Have you had a previous live birth?
(2) If yes, is this child alive now?

Other supplementary information which is useful for analysis is: mother's age; marital status; total children born alive; total children living; birthweight of the newborn child.

The sample size needed for reliable estimates is 1000.

Prevalence of disability

The International Year of the Disabled (1982) aroused a great deal of interest in the prevalence of disability as well as in its prevention and rehabilitation within the community. Unfortunately, very few countries have reliable data on rates of prevalence. A set of Ten Questions has been designed to provide a rapid tool which is also effective across different cultures. The Ten Questions enquire into different types of disability viz. vision, hearing, mental retardation, speech problems, epilepsy and mobility disorders. Validation studies carried out in Bangladesh, Pakistan and Jamaica have so far been encouraging.

Sampling methods

The classical methods of sampling using random tables with or without stratification are too cumbersome for everyday use by health managers. An innovative development has come in the form of cluster sampling. The method involves identifying 30 clusters each of seven individuals from a given area weighted according to the distribution of the population. The method has been extensively used for the evaluation of programmes of immunisation and oral rehydration, and is being adapted for other investigations.

Ethnographic assessment

Modern systems of health care constitute an alien approach to disease in many traditional societies. Utilisation of services may be poor on account of cultural, social, economical and physical constraints. Community surveys using the structured questionnaire may fail to provide the desired information. A focus group discussion, in which the investigator promotes discussion on a checklist of topics will throw up a number of important issues. These may then be included as variables in the final questionnaire. Focus group discussion using client groups from different socio-economic backgrounds, as well as groups of different cadres of health workers, throw light on perceived and actual reasons for non-utilisation of services.

Surveillance and monitoring

Surveillance in the form of systematic collection and evaluation of morbidity and mortality data from the entire catchment area is possible in only a few developing countries. A breakthrough has been in the form of designated sentinal sites. The data from these sites are used as indicators of trends. This approach has been successfully tested in the case of the expanded programme of immunisation and oral rehydration.

Good reporting systems from sentinal sites and laboratories can provide early evidence of new outbreaks for example of diarrhoea, typhoid, meningitis or other diseases. Immediate action for containment of the outbreak can prevent spread and deaths.

Indicators of risk

The mid-upper arm circumference is proving to be a useful tool for rapid survey of a community's nutritional status and for selection of those at risk. First tried as a composite measure together with height in the form of the Quac Stick during the Biafran war, the arm circumference has been used by itself to assess the nutritional status of children under the age of five (for example <13.5cm.), to assess survival chances of newborns with low birth weight (for example <9cm), and for the selection of pregnant women for food supplementation (for example <20cm).

Case-control studies

Comparison between cases and carefully matched controls is a powerful method of identifying social, cultural and biological risk factors for targeting health interventions. A number of refinements have been added to the classical case-control approach to make it as powerful a research tool as cohort studies, but at a fraction of the cost.

No doubt many of these methods will be further developed and refined as more field experience is gained in their application. A desirable development will be a synthesis between social epidemiology and the management sciences, so as to train District Health Managers in the day to day application of the new techniques.

Why does ill-health occur?

When several people are in contact with a pathogen only some people become ill. The same is true at the community level, certain families become ill, and in certain areas more people and families become ill than in others. By looking at these patterns in place and in time some of the factors associated with ill health begin to become apparent.

Interaction of nutrition and infection

It is well recognised that a large proportion of morbidity and mortality arises from the interaction of malnutrition and infection. The malnourished individual cannot muster an adequate immune response to fight infection. Hence even minor infections can spread, causing generalised disease in an organ or a system. Recovery is prolonged and may not be optimal. On the other hand, all infection has an eroding effect on the lean body mass causing loss of nutrients. The nutritional cost of repair tends to be heavy and the local diets are often incapable of meeting all the energy needs of the convalescing patient who may not have fully recovered his appetite. Moreover, the high frequency of infective episodes results in a nutritional deficit from one illness being carried over to the next. This accumulation of deficit after repeated infections is best seen in children in the form of growth failure but also occurs in adults. Hence control of infection and improvement of nutrition needs to be a major objective of the District Health Programme.

A convenient approach to the control of some infections is through immunisation. Both whooping cough and measles are serious illnesses of children with high case fatality rates as well as causing serious and prolonged weight loss. The treatment and rehabilitation costs of tuberculosis and poliomyelitis are high and prevention through vaccination has been shown to be cost-effective.

After measles and whooping cough, diarrhoeal disease is an important aetiological factor in childhood malnutrition. Here early institution of oral rehydration results in diarrhoea causing only a minimal disturbance in the child's health. The illness is less acute and appetite is preserved. Early institution of feeding helps recovery. Provision of facilities for oral rehydration at the village level has resulted in reduced morbidity and mortality as well as an understanding of the fluid and nutritional needs of the patients.

A mistake in the past has been the failure to recognise the important role of infection in the aetiology of malnutrition. Nutritional status had come to be equated with nutritional intake alone and hence animal protein was awarded an important place. Energy intake and infection received only passing mentions. Now poor energy intake has been rightly recognised as being significant in the aetiology of malnutrition. And so also is infection. Hence the provision of adequate food and energy as well as control of infection must be integrated in all nutrition programmes.

Nutritional problems in developing countries

The major nutritional problems of protein-calorie deficiency, anaemia, blinding malnutrition, specific vitamin deficiencies and endemic goitre are now recognised. Of these, protein-calorie deficiency and anaemia are the most prevalent and have far-reaching effects on the health status of the community. When food is scarce, almost the entire family suffers from deprivation, but in general the effects are more devastating in those who are biologically most in need of food viz. the growing child, the pregnant and lactating women, and the convalescent. Some traditional diets have high roughage content and consequently low calorie density. This can cause problems with weaning foods. For example, some traditional weaning foods provide 1 kcal/g. of food compared to breast milk which has the energy density of 6 kcal/g. dry matter. The beginning of most childhood malnutrition can be traced to the weaning period. The slowing of growth in the first year of life accounts for 91 per cent of the deficit in body weight and 98 per cent of the deficit in length seen at the age of three. The low calorie density of some traditional food is also a limiting factor during convalescence when rapid catch-up and restoration of lean body mass is essential. But anorexia and weakness prevent the sick child from eating adequate amounts. A nourishing diet with adequate calorie density is also especially important at this period. One way of increasing the calorie density of foods is to add fats and edible oils in cooking. Red palm oil, coconut, cotton seed oil, ground nuts, soya and other sources of high energy foods are important nutritional resources which have been hitherto ignored because of preoccupation with animal protein.

Growing of food and its consumption is one of the most basic human activities. It is influenced by the laws of demand and supply as well as distribution, as is the case with every productive activity. Distribution of foods between and within countries, between rural and urban areas and within the family is sometimes as important as biological considerations of composition, methods of cooking, and so on. Rural health programmes need to take this aspect of distribution of an important resource into account. In most cases intervention is needed at the community level in addition to that at the level of the family and the individual.

Role of parasitic diseases

Malaria is ubiquitous in all countries of the tropics and the sub-tropics. In the early months of life the infant is relatively well protected because of the transplacentally-derived maternal antibodies. But as their effects wane, the infant and the young child become increasingly susceptible. This is therefore the age during which the main brunt of malaria is felt. In countries like Sri Lanka and Mauritius where effective malaria control programmes have been carried out, child mortality has been reduced by as much as half in some instances.

Another group in whom malaria is a serious problem are the pregnant women. Due to changes in the immune response caused by the hormonal changes in pregnancy, many women become increasingly vulnerable to malaria. Anaemia of haemolytic type is common and interferes with placental growth and function. Moreover, heavy placental infection by malaria parasites is also common so that low birth weight is a common complication of malaria. Infant mortality is closely related to the birth weight of the baby, the influence of which on survival can be identified up to the age of six months and beyond. Thus, malaria is an important factor in infant and child mortality.

In the adult, the main effects of malaria are in the form of chronic anaemia, lack of energy and vitality, together with a feeling of being unwell and increased susceptibility to infection because of compromised body reserves.

Other parasites

Similar reasoning applies to the effects of other parasitic diseases. Their effects depend upon the age and nutritional status of the host, the size of the parasitic load and environmental influences. Thus, even though a considerable number in a defined community carry a parasitic load, only a proportion suffer disease. From the disease point of view, the important parasites are *Entamoeba histolytica, Giardia lamblia, Ascaris lumbricoides, Ankylostome duodenale*, together with *Necator americanus, Strongyloides stercoralis*, and schistosomiasis. Infection is widespread, and so by comparison they constitute a large proportion of disease of parasitic origin seen. Many of the parasites have a complex life cycle utilising one or more intermediate hosts and their control will often involve control of such intermediate hosts as well as dealing with the reservoir in the community.

Determinants of disease and the physical, social and cultural environment of the individual

Mortality and morbidity figures convey only factual information. They cannot tell about the causation of disease and its determinants which often lie within the physical and socio-cultural environments and within the life-styles of the people. If health services are expected to help reduce the incidence of disease instead of being just palliative, then it is essential that the determinants of illness in the community are identified. Information obtained from morbidity and mortality statistics needs to be supplemented with other studies or surveys and particularly observation. In addition, KAP (knowledge, attitude, practice) studies, nutrition surveys, household surveys, and also agricultural profiles can be useful. Such studies tell us about what people do, when and why. It is only through creating a sufficient data bank and reliable system of health information relating to the community that health programmes and activities become relevant.

A great deal of disease in developing countries is environmentally determined. Even though the physical environment is obvious like vegetation and its insect breeding sites, housing, neighbourhood, sanitation, geography, seasonal climate and so on, the social and cultural environments are equally important but less obvious. Man is very much the product of the immediate society which he in turn influences by contributing to its social, economic, cultural and political life. Such interactions occur both at the individual level and at the family level, the latter being the unit of society. Most countries of the Third World are in a process of rapid transition so that a process of change is going through many of the socio-economic and political-cultural institutions of these countries. For many people life-style is changing, bringing new habits and new ways of thinking. Social and cultural currents run through all communities and societies. The concept of disease causation, the divining of illness, the selection of the provider of care, the customs and practices related to child-bearing and child-rearing are all integral parts of the provision of services. A good knowledge of local practices and socio-political trends is essential for providing the kind of service that people will perceive as stemming from national roots rather than a foreign concept.

Underlying many of the environmental determinants of health are the common elements of poverty and inequality. In all societies inequalities exist. There are inequalities of resources, privileges, opportunities, education and mobility. Naturally, inequality in one area engenders inequality in a related area, resulting in a class structure. It is said that in any social group with an existing inequality, any new resource or service will be shared out in accordance with the inequality. Those who are well-off will gain more than those who are badly off, unless special steps are instituted to avoid such an imbalance at the time of planning the service. Several corollaries arise from this aphorism. The most important, from the point of view of planning health care, is that when great wealth exists in a society, there is also great poverty, and often the former contributes to the latter. More often than not, the health professionals are also members of the upper social class and unknowingly part of the existing inequality. In almost all countries, therefore, the Law of Inverse Care applies so that those in greatest need of services get the least. Hence, in designing and implementing health programmes there is a great need to cater for the needs of those who are excluded from, and marginal to, the mainstream of development.

WHAT IS WRONG WITH THE EXISTING HEALTH SERVICES?

Health services do not always work perfectly but when a new plan is drawn up people may forget this. Planners may assume that with a new plan all the problems have been taken care of. In fact, if the existing problems have not been recognised and faced, they will merely be perpetuated into the new system.

Many developing countries are still in a state of transition from colonial rule to independent self-government. The seeds of the administrative and social services were sown during the colonial era and were essentially based on a Western model of planning. Even though many scientific and social developments occurred during the colonial era, the services tended to become 'institutionalised' and never 'popular'. In the case of health, for example, at the time of independence a framework of Regional and District Hospitals existed and the concept of the Health Centre was just about gaining ground. After independence an important influence on planning has been the professional one. Many of the leading professionals had received their specialist training abroad and were naturally deeply committed to the establishment and perpetuation of their own speciality. The result has been a continuing emphasis on 'curative' care based on hospitals and discrediting of the promotive approach in rural areas. This disparity between curative as against preventive/promotive medicine also permeates medical education which is largely based on curricula developed by medical schools in the West. In the case of the latter, public health and preventive medicine are often taught as postgraduate disciplines whereas in developing countries the need for teaching those subjects in the early years is obvious. Moreover, the practice of public health in a predominantly urban, literate, and technologically advanced society is so different compared to that in a predominantly peasant society though the principles may be the same. The District Medical Officer is a product of this system of health planning and education and will have to make continuing efforts to learn and match his skills to the problems in rural and peri-urban areas.

Are the services coping?

This question should be continually in the minds of all those who are responsible for providing services. The creation of health facilities and services does not in itself make a health programme. The activities of some of the services may be totally irrelevant to the health needs of the people.

Coverage (Are the services adequate?)

The main issue is that of coverage, and this gives rise to several subsidiary issues in the politics and sociology of health.

Several studies in different countries show that the large urban hospital serves the need of the educated élite but fails to meet the needs of the rural people as well as those of the 'fringe people' living in urban slums, shanty towns and inner city areas. The present curative system of health care has been correctly described as 'importing from abroad yesterday's solutions for tomorrow's problems'. A new approach and thinking are necessary for creating appropriate services to meet the needs of family health in developing countries.

With regard to providing coverage, *geographic coverage* of all areas of settlement in a District is important. In most countries provision exists for one rural Health Centre for every 100 000 population and a sub-centre for every 10 000. The geographic location of these health institutions is important since in practice most rural health units are accessible to people within a five-mile radius. They cannot be expected to provide effective coverage beyond this area. Whereas the sick can be expected to travel long distances for relief of pain or symptoms, services for the healthy must be delivered to them as near to their homes as possible. This is especially so with regard to services for women and children. The pregnant woman cannot be expected to make a round trip of more than six to eight miles on foot, often with a toddler in tow. Hence antenatal services and under-fives' clinics need to be organised at a greater number of places and not restricted to only Health Centres and sub-centres. Ideally these clinics should exist for every large village or cluster of hamlets.

Still on the subject of coverage, the frequency with which the antenatal and under-fives' clinics are organised will enable the people to make greater use of them. A clinic that operates on a weekly basis provides more effective coverage than the one operating on a monthly basis, and a daily clinic will be more effective than a weekly one. This ideal may not always be possible and a compromise formula may be needed in which clinics operate in rotation in such a way that on any given day a clinic is available within a four- to five-mile radius of every home in the District. Outreach or satellite clinics not only help to extend coverage, but enable the Health Team of a Centre to visit front-line health workers regularly and help to train them and upgrade their work.

Measuring coverage

Coverage is determined by the presence of a number of constraining factors. A framework for the measurement of coverage can be developed by putting together these factors as follows:

(1) Availability of resources as measured by the proportion of time in the year during which the resources needed for implementing a given

intervention were available (for example availability of: vaccines; oral rehydration packets; growth charts; antenatal cards; essential drugs, and so on).
(2) Geographical accessibility of services as measured by the per cent of target population living sufficiently close (for example half an hour's walking distance) to service delivery points, and to referral and back-up services. Accessibility can vary depending on the season, condition of the roads, availability of transport, costs of transportation, and so on.
(3) Utilisation of services by the target population as measured by the actual contact between the target population and the services. The measures of utilisation can be the number of prenatal and postnatal contacts; uptake of immunisation or oral rehydration; and by attendance for individual illnesses.
(4) Compliance, as measured by finishing a full course of, for example: antibiotics; immunisation; antenatal care, and so on.
(5) Effectiveness of coverage as measured by the quality of care provided (for example standardised treatment for a given illness; close adherence to the requirements of the cold chain and vaccination techniques; follow-up of those 'at risk' according to a standardised protocol etc).

Using such a framework a matrix of coverage with major PHC activities in the catchment area of each health facility can be prepared. For example, coverage with the EPI (expanded programme of immunisation), ORT (oral rehydration therapy) ARI (acute respiratory infections), safe motherhood, promotion of breastfeeding, family planning, and similar other programmes can be assessed from time to time. Such exercises are best done jointly with the community. Then problems linked to the supply of services like availability, accessibility and effectiveness of coverage can be addressed by the health team, and problems associated with utilisation and compliance can be attended to by the community.

Are Mother and Child Health (MCH) services being provided?

In most developing countries mothers and children constitute up to two-thirds of the population of an average District. They also constitute the biologically vulnerable groups. Hence Mother and Child Health (MCH) services constitute an important area of health care. Deficiencies or absence of any of the following basic health activities necessary for providing coverage in MCH shows problems in existing services which need to be rectified before trying to extend services further. For example, if midwives currently do not recognise high risk mothers, how can they be expected to train Traditional Birth Attendants (TBAs) to do so?

Basic activities in Mother and Child Health (MCH)

(1) Screening of expectant mothers. Identification of those at risk or with abnormalities and their referral for more expert care.
(2) Assistance during delivery and puerperium.
(3) Growth monitoring, regular health surveillance and immunisation of children. Identification of high risk families and further care of such families.
(4) Simple recording of health events in individual children, using a weight chart. Data collection and evaluation.
(5) Health education emphasising nutrition, child-rearing, immunisation and fertility problems.
(6) Providing information on community health problems to other agencies and workers in the area and also to the community itself.
(7) Counselling and assistance with family planning.
(8) Distribution of simple medicines, food supplements and contraceptives.
(9) The recognition and primary management of the most common diseases in the area.
(10) Participation in the control of the communicable diseases through immunisation, through diagnosis and treatment of index cases as in tuberculosis, or through treatment of a reservoir of disease as in mass de-worming.
(11) Liaison with community development, agricultural extension, education and other similar services in the area.

On these core MCH activities further services may be added, depending upon local needs. For example, nutrition rehabilitation centres, mothercraft classes, demonstration vegetable gardens, and so on. Adult literacy, especially female literacy, has an important bearing on the health of the family and several countries have integrated programmes of adult literacy with MCH services. Based on recent experience in Sri Lanka and Kerala, it has been suggested that for every one year of average schooling of girls, a reduction of 10 percentage points in infant mortality can be reasonably expected.

Are people utilising the services?

Many people are reluctant to make use of medical services. This was thought to be largely due to ignorance or cultural beliefs, but other factors play a part too. For example, Health Centres and sub-centres stand out as different from other constructions and dwellings in rural areas. Their style of construction, roofing, and the finish are different and most are fenced in. The health workers inside them are in uniforms and usually are people who were born in some other part of the country. They have a different life style and rarely do they participate in the social and cultural life of the village. Naturally the villagers are reluctant to 'intrude' on them. Attendances have been improved in many cases by 'localisation' of institutions through renting buildings rather than putting up new ones. Use of auxiliaries who are local residents and of village health workers as well as 'volunteers' from the local population also helps to remove local fears.

Community involvement at all stages – planning, implementation and evaluation – has an important effect on utilisation of health services. Involving the community demands continuing dialogue with the community and the establishment of village health committees. It is in these committees that most of the administrative and managerial problems can be aired. The health committee brings its own rewards usually through making local resources available, for example, a local building for holding clinics, selection of village health workers and their remuneration, volunteers and supervisors for health campaigns and so on.

Health surveillance of vulnerable groups requires more administrative and managerial skills than specialised medical knowledge. Many non-medical individuals in the community who are literate and can communicate well are able to help in several ways. In the clinics they can assist with weighing and recording and in being the first contact with parents. The village teacher or social worker can help with home visiting and counselling. School children, by their involvement with school gardens and preparation of school meals, are a medium through which village schools can contribute to better eating habits. Indigenous midwives and even traditional practitioners have been enrolled to help in some cases. All these groups can generate a community interest and help improve attendances. Such use of available local resources represents an important element of the Primary Health Care concept. It enables community health programmes to be put into effect without undue delay.

Is the 'at-risk' concept being used in provision of health services?

The MCH team needs to select individuals and families 'at risk' of ill health for special care besides providing regular health surveillance of mothers and children. For example, the antenatal clinics are meant chiefly to provide the basis for healthy motherhood and not to diagnose obscure disease. According to criteria established by a simple study of local records, mothers may be selected for special care from among those attending the clinics. There will be a *primary selection* depending upon the obstetric and personal history of the mother, for example, very young, poor obstetric history, past history of complicated or instrumental delivery, history of stillbirths and neonatal deaths, failure of lactation with previous children, maternal height of under 5 ft (150 cm) and so on; and there will be a *secondary selection* depending upon disease processes occurring during the course of the pregnancy, for example, hypertension, toxaemia, anaemia, or social problems at home. In the same manner, the under-fives' clinic is meant for growth and nutrition supervision of the children and prevention of illnesses such as whooping cough, diphtheria, tetanus, poliomyelitis, measles and tuberculosis. From these clinics children may be selected for special care according to the following criteria:

(1) Broken homes, death of one parent or lack of parental care.
(2) Onset of another pregnancy in the mother.
(3) Failure of lactation.
(4) Multiple pregnancy, so that available breast milk is not enough.
(5) Low birth weight.
(6) Death of siblings.
(7) Presence of chronic illness in the family.

The special care provided may consist of more frequent visits to the clinic, home visiting, nutrition rehabilitation and health education, supply of food supplements, family planning or referral to other social services in the community.

Is there adequate quality of care? (See also Standards page 178)

The type and quality of care provided can influence the determinants of disease in the community. For example, when all Health Centres and sub-centres become just the extensions of the hospital out-patient departments they can have no impact on the health of the community. And this in fact has been the case with many health institutions. Small scale time/motion studies will indicate how the time of different health workers is utilised, and appropriate modifications can then be instituted. Such a study of the auxiliary nurse midwife in health centres in India showed that she spends most of her time in assisting for curative care and very little time on preventive/promotive care. A similar time/motion study of time taken by community clinic attendants to diagnose and write prescriptions gave an indication of the time taken up for treating common complaints (see table 2.14).

Related to the question of the quality of care is a subsidiary question – Are the health services relevant to the health problems of the community? What modifications and changes can be introduced to help the services deal with the major health problems more effectively? For example, worm infestation may be a major problem. The vermifuges dispensed at the sub-centre, however effective for the individual case, may have no effect on the 'reservoir' in the community. Advising building and using latrines can affect the problem only gradually. On the other hand, a mass campaign for de-worming can produce immediate results and help to build confidence for the construction of latrines.

Table 2.14 **Consultation time for common conditions observed during a period of 4 days in 15 community clinics in Ghana.**

Time (minutes)	Cough	No. of Cases Diarrhoea	Fever	Total (%)
5	9	15	15	39 (19.5%)
5– 9	33	44	38	115 (57.5%)
10–14	18	13	11	42 (21.0%)
15+	3	1	–	4 (2.0%)
Average time per patient (minutes)	8.6	6.5	6.3	7.1

Source: Amonoo-Lartson, R. et al. Soc. Sci. Med. (1981) 15A, 735–41.

Similarly, in the case of tuberculosis, case-finding by sputum examination and treatment may have a greater impact on the prevalence of the disease than mass miniature radiography.

Is staff morale high?

An important element in the quality of care is the general morale of the staff. Regular meetings with the staff, help with administrative and professional difficulties, involvement in decision-making, a regular supply of medicines and professional support helps to improve the quality of the care they provide. When health workers in remote rural areas feel abandoned, on account of infrequent visits, irregular supplies and no opportunity to improve their knowledge, morale sags and the quality of the work suffers (see also chapter 5, pages 199–208).

Do staff interact?

If the health workers in the District are to become a proper Health Team then there is need for them to get to know each other and each other's difficulties. Regular meetings, refresher courses and seminars go a long way to building up a team spirit. Also, if the District Health Service is to become a viable system then, like any other system, it needs two-way traffic. There should be a flow of professional and administrative support from the centre to the periphery and regular analysis of health problems and effectiveness of programmes in the opposite direction. The greater the interchange and flow of information and ideas within the system, the more sensitive it becomes to the health needs of the community.

At present the Health Centres and sub-centres largely function as mere conveyor belts feeding 'interesting' clinical material to the specialists in the hospital. Such a system of health care can hardly ever expect to change the health situation. A new attitude in health care delivery is essential, such that the action shifts to the peripheral units and to the front-line health workers. In such an approach the role of the District Hospital and its medical officers is more in the nature of being supportive, managerial and administrative. Medical, professional and community resources can then be harnessed to deal with determinants of disease in innovative ways to bring about change.

Is there regular health services evaluation?

In all health work, regular and continuing evaluation is an important aspect of the activity. It enables the health team to decide upon priorities, to select areas where intensive effort is needed, to identify problems and to choose those approaches which appear most likely to be fruitful.

The community should be actively involved in the evaluation and be represented at all meetings where evaluation is being discussed. Similarly, various groups in the community can be of assistance in gathering data for evaluation.

Besides being essential for health planning, regular evaluation has a further advantage. It provides a basis for establishing a dialogue with the community through which the interest of the people can be maintained and their cooperation obtained for further development of health activities.

The evaluation component must be incorporated into a health programme right from the start, and must be designed to yield information for decision-making both at the operational and the planning levels. This calls for a data collection system which is simple and relevant to the successful operation of the programme. The purpose of evaluation is to improve staff performance and not as a punitive measure.

Microplanning following evaluation of coverage

When results of monitoring of coverage pin-point bottlenecks a plan of action can be jointly evolved with the community. Such a process is to be looked upon as a dynamic one in which objectives, corrective actions and resources are continuously adjusted in response to the results of the evaluation. Microplanning is thus a planning process carried out by and for each health facility within the overall District Health Plan. It provides the fine tuning necessary for solving operational problems at each service delivery unit. It can identify the need for staff training, for better logistics, for better organisation of work, for closer involvement of the community, and so on.

Following each evaluation microplanning addresses the questions listed on the following page:

(1) What corrective actions are needed?
(2) Who will be the people responsible for carrying them out and who will be the target groups?
(3) Where and when will the activities take place?
(4) What resources are needed?

IDENTIFYING LOCAL RESOURCES

To identify health care resources in the District, the same five questions can be asked – who? what? where? when? why? In any given situation resources can be classified under the three major headings of: people, time and money, in that order of importance. In many health activities undue emphasis is placed upon money, so much so that its availability has become the only deciding factor for the commencing of any programme. Another less known aspect of resource is that in many cases the resource needs to be developed and strengthened. In the absence of such a careful 'tending' of resource it may often degenerate into a liability. For example, a healthy population full of vitality and with community cohesion is an important resource for development. But a sick population divided amongst itself and apathetic is a liability. The same principles apply to other forms of resource also. Money well spent on effective health programmes based on sound epidemiological data can provide immediate results. On the other hand, if it is tied up in impressive buildings then annual recurrent maintenance costs alone will be a crippling burden for years to come. The resource of time well utilised is an investment in the future. But if time has been wasted in getting projects off the ground, then not only can it never be recovered, but during that period the population (and the problem) may have grown larger.

Who is providing health care? Who do people go to for advice? Where? When? At what cost?

Who gives health care and advice?

Many different types of people provide advice and health care. If a child is ill, it is often a neighbour or grandparent who is the 'nearest source of help' and who will be the first to discuss with the mother what care is needed. After consulting near relatives or neighbours, the next type of person approached about sickness in a family is likely to depend on what the family thinks are the causes of ill health and which services they think may be able to help (see figure 2.3). In many societies the next source of advice and care is the drug seller or chemist or traditional healer. Spiritual healers may be sought to use their skills for specific problems. Other problems may be taken to the Health Centre (see figure 2.4).

District health needs

TYPES OF PRACTITIONER	BACKGROUND AND TRAINING	PROPORTION % OF ALL PRACTITIONERS	DIARRHOEA	FEVER	RHEUMATISM	RESPIRATORY INFECTION	WORMS	JAUNDICE	FRACTURES	SNAKE BITE	HEADACHE
ALLOPATHS	High School 1-4 years apprenticeship	15%	◼	◼	◼	◼	◼		□		
HOMEOPATHS	Follows a school that uses minute quantities of medicine	3.3%	◼	◼	★	◼	★				
KOBIRAJ	Ayurvedic training in herbs, minerals and diet	15.3%	★	★	◼			★	◼		★
TOTKA	Use Ayurvedic, Yunani and Shamanistic methods	60%	★	★	◼			◼	◼	◼	◼
OTHERS	Yunani and Fakirs	6%	★	★	★					★	★

★ over 10% of practitioners considered the disease their speciality

□ over 10% of clients would visit the practitioner for this disease

Figure 2.3 Specialisation amongst traditional practitioners: Bangladesh

Source: Glimpse 1980.

District Health Care

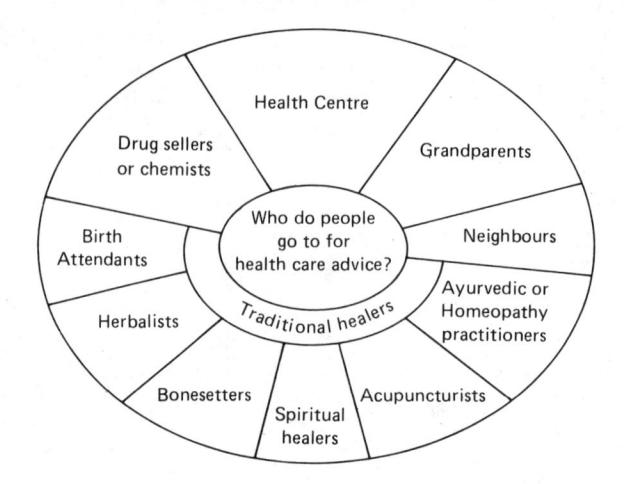

Figure 2.4 Who do people go to for health care advice?

The preference of parents in the Punjab for the source of advice and treatment for some childhood disease (N = 60)

Folk practitioner   Veds or Hakims   All other practitioners   Health Centre

Small Pox / Chicken Pox / Measles: 100%

Tetanus Neonatorum: 60%, 23%, 11%, 7%

Pneumonia: 87%, 13%

Source of health care in rural North India

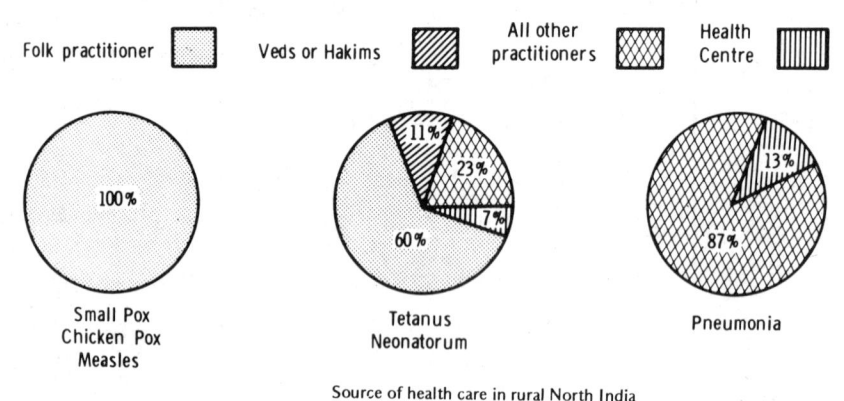

Figure 2.5 Who people consult about specific health problems (Punjab, India, 1972)

Source: Kaher *et al.* 1972.

There is growing evidence of the use of this wide diversity of sources of information. In one small-scale study in Newcastle upon Tyne, mothers in a low socio-economic neighbourhood were asked what action they took when their child was last ill. Nearly all the mothers had discussed the child with a grandparent (or sometimes a neighbour) to decide whether the child was ill enough to warrant going to the family doctor. In a larger UK study it was found that many people go to a chemist's shop to ask for help and purchase lotions and medicaments to try to solve their problems. Only certain problems are taken to the formal health services. The same pattern is found in other countries too. Figure 2.5 shows who people consulted in the Punjab for specific conditions. Skin rashes and spots of all types and also tetanus were likely to be taken to folk practitioners, pneumonia was likely to be taken to private practitioners or occasionally to the Health Centre.

In parts of Africa people prefer to consult a bonesetter for a compound fracture rather than go to the hospital. Where the causes of ill health are thought to be social and psychological, psychic healers may be consulted. Chronic illness is increasingly felt to be inadequately dealt with by the Western health system, and so when those suffering from chronic problems have tried the health care system, they may well try the many methods of alternative medicine. Many others feel the need to consult a health care provider who they feel understands their problems and background. Perhaps this is why hakims are in demand in West London and also why some Western women prefer to go to 'women's' hospitals or health care institutions.

Drugs are increasingly easy to obtain in many countries, from travelling drug sellers setting up stalls in markets and via chemists or drug stores. Many people are delighted to be able to buy what they need – family planning supplies, aspirin for headaches, and chloroquine for malaria. All these sources of health care are often far easier for most people to use than the health care system based in clinics and units often much further away.

Doctors, nurses, midwives, medical assistants, auxiliary nurses and environmental health personnel will all be providing health care in the District, some working for the government services, some for non-governmental organisations and others working privately. In addition to government health workers it is also useful to find out how many private health workers are in practice and also the number of non-government health units such as mission stations. These are resources with whom the District Manager can often work in close collaboration.

In addition to *existing* staff some information is needed on future staff. One scheme for doing this is shown in figure 2.6. Such an analysis can lead to figures such as the following graph (figure 2.7) for Ghana. It shows a huge increase in enrolled nurses if projected to 1990 and only a very small increase in community nurses. When this was realised there was a policy change because it was community nurses who were needed for the future and not hospital enrolled nurses. Only by such analysis can underlying trends be recognised before difficulties arise.

Experience in several countries, chiefly in tropical Africa, but also in Papua New Guinea, Malaysia and Indonesia, has demonstrated the key role of the medical auxiliary for health work in rural areas. Yet the 1960s and 1970s were notable for the debate and the professional scepticism about the capabilities of such a person. The debate is now settled largely in favour of the auxiliary. A great boost to the concept came from the establishment of training

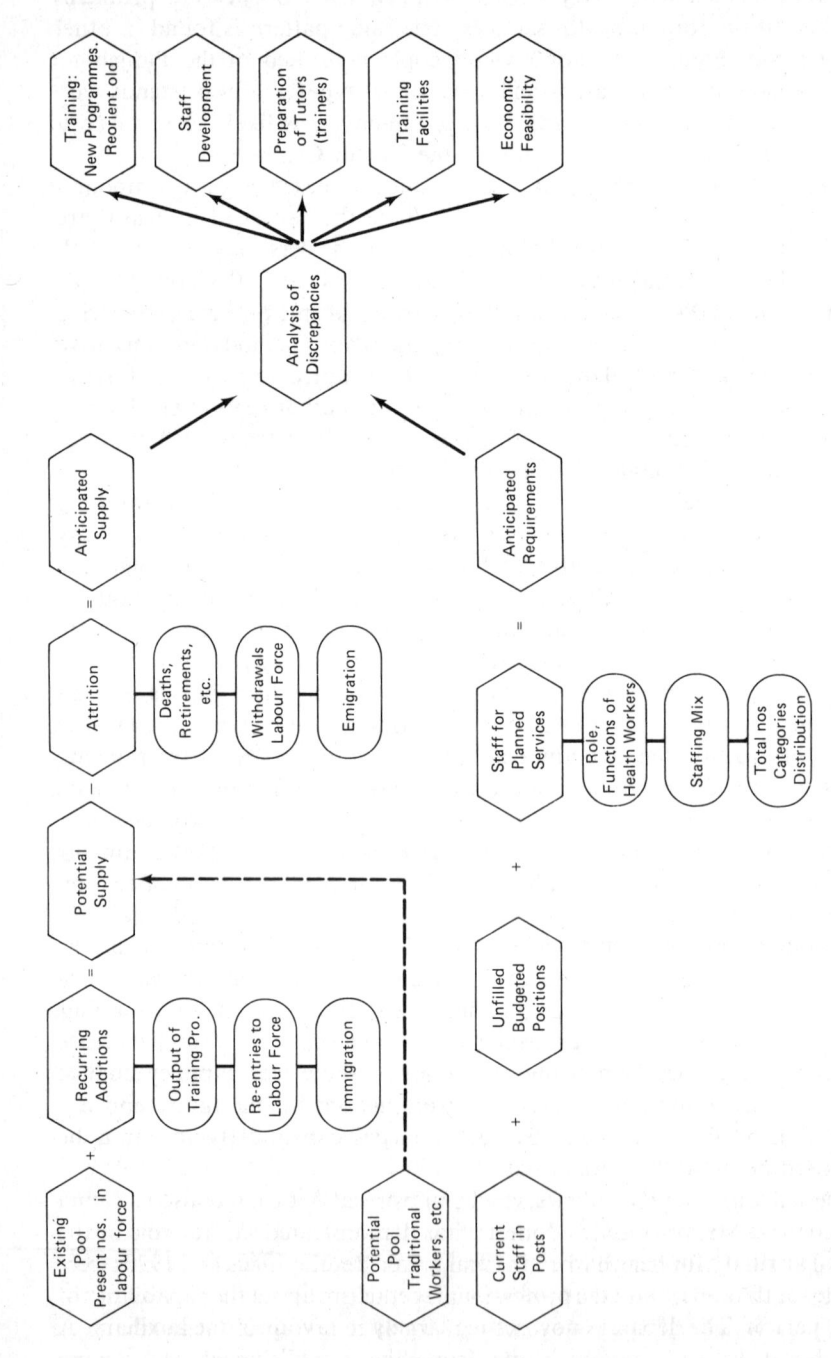

Figure 2.6 Model for Health Manpower Planning

Source: National Health Planning Unit, Project Team Spec./Human Resources, March 1978. Ghana.

District health needs 69

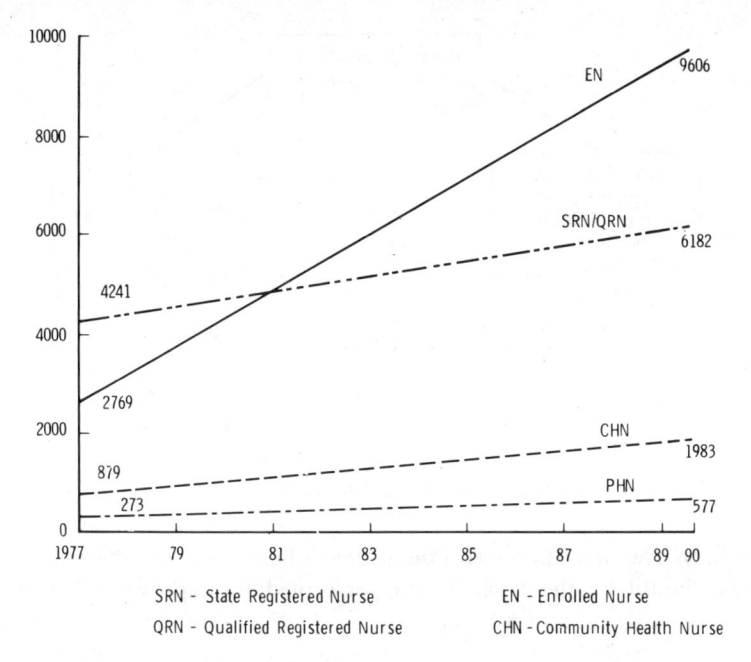

Figure 2.7 Projected supply of nursing personnel in active work in Ghana 1978–1990.

Source: National Health Planning Unit, 1978.

programmes for the 'Medex' and the 'physician's assistant' in the United States. The debate in the present decade is about the part-time village health worker. This cadre is being trained and established as part of a national policy in many countries and in response to the call by the World Health Organization for Primary Health Care for all by the year 2000. For the doctor responsible for the health care of a District it will be more pragmatic to look upon the auxiliary, the village health worker, and the birth attendant as health resources, and to develop the quality of this resource through training and professional support as well as good administration instead of undermining their confidence through undue criticism.

National health plans are normally put together at the national level and the good ones make provision for manpower requirements on a national scale. District Health Teams will then have the responsibility of recruiting and training front-line workers for carrying out health activities at the periphery. Professional and the higher auxiliary grades are usually recruited at the National level and trained at National or Regional health training institutions. At the District level, health manpower development must be more specific and task-orientated. In other words, the problems to be dealt with must be clearly defined, the tasks to be performed clearly spelt out and defined in detail and the staff required to perform them recruited and trained (see

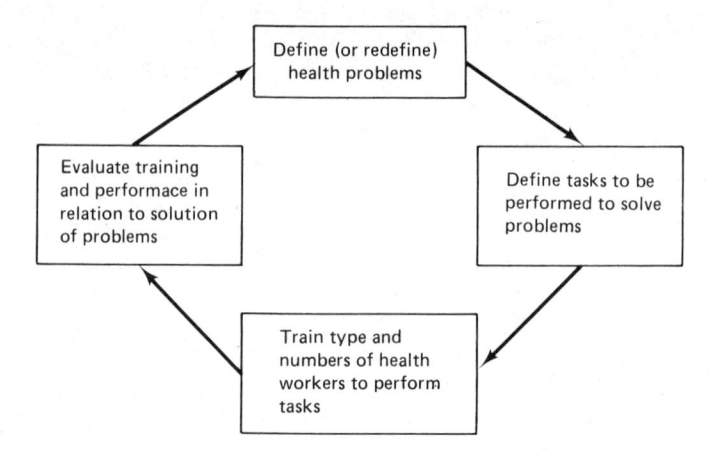

```
                    ┌─────────────────┐
                    │ Define (or redefine) │
                    │ health problems │
                    └─────────────────┘
                   ↗                   ↘
┌─────────────────┐                   ┌─────────────────┐
│ Evaluate training│                   │ Define tasks to be│
│ and performace in│                   │ performed to solve│
│ relation to solution│                │ problems        │
│ of problems     │                   │                 │
└─────────────────┘                   └─────────────────┘
                   ↖                   ↙
                    ┌─────────────────┐
                    │ Train type and  │
                    │ numbers of health│
                    │ workers to perform│
                    │ tasks           │
                    └─────────────────┘
```

Figure 2.8   Assessing future staffing requirements

figure 2.8). The more peripheral the cadre of health worker, the more clearly defined should be the tasks to be performed to improve efficiency and effectiveness. Regular supervision and evaluation should be carried out to provide the background for a continuing programme of retraining and reorientation.

Where is health care provided and where do people come from to use it?

(a) *Where is health care being provided?*   Mapping Government and voluntary agency health facilities is most useful. Ordnance maps are available in most countries, and show the major towns and villages as well as roads, railways, rivers and other geographical boundaries. On such a map the District health facilities can be identified using symbols or coloured pins.

The district hospitals.   The District Hospital(s) plays a key role in a District's health care system if the latter is properly organised and well managed. Without the back-up facilities of the hospital, cases of complicated or advanced illness being referred as a result of increased health coverage cannot receive the special care they need, and the credibility of the health programme may suffer as a result. If traditional birth attendants are trained to screen for obstetric difficulties during the antenatal period, then facilities must be provided to deal with the referrals. This is good management. Emphasising preventive care does not necessarily mean that hospital care is to be neglected. In fact prompt treatment will often reduce the severity of illness. On the other hand, emphasis on preventive care is essential because it has not received importance in the past and because many illnesses are preventable. A good way of planning for the hospital needs of District health services is to compile a list of common medical, surgical and obstetric emergencies treated at the hospital. A review of admission registers and a few questions to hospital-based doctors and nurses will help generate such a list. Such an exercise will indicate the basic equipment and facilities needed at the District Hospital. In many places hospital outpatients and casualty departments provide the first point of contact for people seeking health care. Primary care provided only at the hospital is extremely expensive and an abuse of resource. It also leads to swamping of services so that the hospital workers are unable to look beyond the immediate problem of dealing with the crowds. They have no time left to see what is happening in the community.

(b) *Where do people come from to use health services?* A rough idea of the areas served by a health unit can be obtained by recording the names of patients' villages as written in the out-patient register for every tenth attendant in a certain time, perhaps one year. This was done for the period 1 November 1968 until 31 October 1969 at Nkhata Bay District Hospital, Malawi. It was found that 47 per cent of out-patients came from within 1 mile (1.6 km) and 61

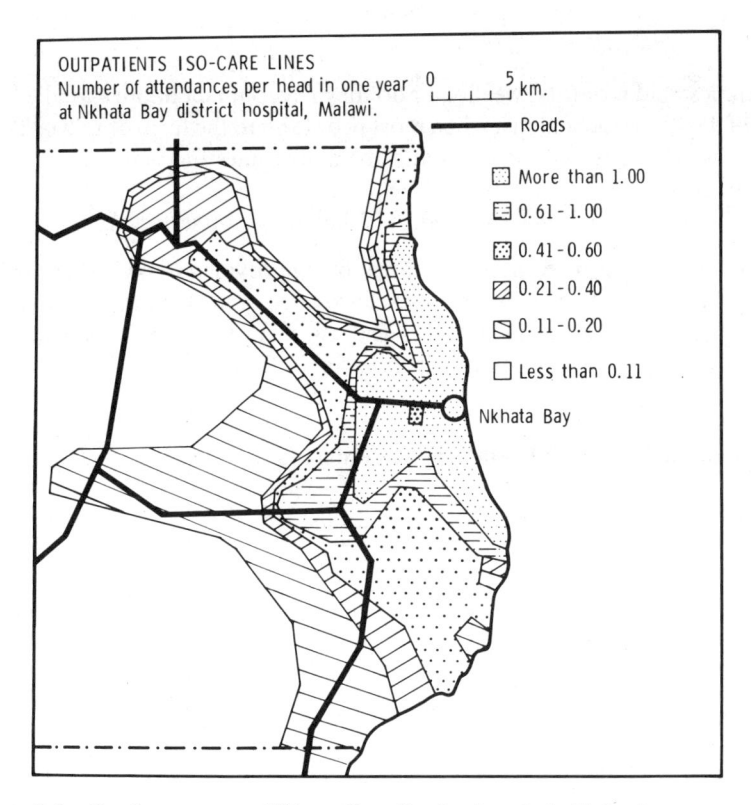

Figure 2.9 Catchment area, Nkhata Bay district hospital, Malawi

Source: de Winter, E. R. (?1974) *Health services of a district hospital in Malawi*, page 110, Van Gorcum and Comp., N. V. Hubrecht Janssen Fund and Schiffner Fund, Amsterdam.

per cent from within 2 miles (3.2 km), 79 per cent came from within 5 miles (8 km) and 91 per cent within 10 miles (16 km). This decline in attendance with increasing distance is well known. In Tanzania it was found that up to 90 per cent of the patients come from within a radius of 5 miles.

However, this 'concentric circle' model of average attendance rates for all villages does not tell the whole story, since it is so dependent on population density. People obviously will travel further where routes are easier and transport readily available. One way of identifying the actual geographic range of a health unit is to calculate the percentage of the population seen for first attendance at an out-patient clinic; lines around areas with similar attendance have been called 'iso-care' lines (King, 1966). These are shown for the Nkhata Bay District Hospital in figure 2.9. The lines appear to follow the roads and the lakeshore very clearly. This confirms that communication (by road, or canoe on the lake) is very important for people when deciding whether or not to go to a health unit.

When is health care provided?

Are clinics held monthly, weekly, or do they never in fact happen at all? Is the time of day they are held just when most people go to the farm or to work? Are the times geared to market days and the availability of transport?

What health care is being provided?   (What range of services?)

One way of measuring what health care is being provided is to visit each health unit and find out when what services are provided. Another procedure is to go to traditional healers and their clients and ask them what conditions they treat or who they go to for specific help (see figure 2.3).

People other than health workers as resources

In every District a nucleus of professionals and para-professionals exists as part of the administrative and civil services. For example, in the health services there are District level specialists, general duty medical officers, nursing staff, medical auxiliaries and para-medical people. They have their counterparts in agriculture, water development, community development, road and communications, education and several other specialities. They form the technical-professional nucleus for activities within the District. Many of them also act as technical advisers to their counterparts in administration. Regular meetings of all the senior administrative and technical officers in a District are essential to integrate activities, formulate new projects and evaluate existing ones. For such meetings a right attitude to planning is important. In the words of Julius Nyerere, the function of the experts is to help people achieve the objectives they (the people) have decided upon and not to decide projects in closed meetings for them. However sound a project may be technically, the people will not make full use of it if they feel that it has been 'imposed' upon them. Involvement of people in decision-making at the political/administrative level is easier in countries with a policy of decentralisation. The idea is to shift decision-making to the District and Regional (Provincial) level instead of letting national and international experts in the capital city decide.

Away from the District administrative centre, technical help can still be found in the form of retired professionals and artisans. Many of them may be happy to do part-time work to augment their pensions. Thus builders, draughtsmen, plumbers, carpenters and many such artisans can be utilised for maintenance or extensions to buildings or for small-scale constructions. Protection and maintenance of wells and water tanks, or putting up new ones can be carried out with the help of such people who are not only cheaper, but also know the local conditions better than city-based specialists. In the Jamkhed project a local artist has been utilised to produce all the health education material used by village health workers. The drawings carry a good likeness to local features, clothes and habitats and are more effective compared to the bland material being put out by the national centre for health education.

Non-professional literate people can be identified in rural areas and encouraged to join as volunteer group leaders. Thus, plantations, mining concerns and small industries in many rural areas employ administrative and clerical staff. Shopkeepers and other self-employed people may be literate or possess skills which can be pooled together. Often their wives and other women folk may be literate or more enlightened than the general population. Such individuals can be encouraged to form volunteer groups for literacy, cooking, mothercraft or sewing classes or for organising child-minding and play groups. Such group leaders also need periodic training in becoming successful organisers and leaders and the District Health Team must take a special interest in organising such training.

Rural life is no doubt one of hardship but not necessarily one of failure against heavy odds. There are always some individuals who make a success of it. The successful farmer, the grain merchant, the successful cattle trader, as well as good parents, can be identified and persuaded to 'teach' others. Often these people have lively minds ready to adopt new ideas and equally ready to communicate new ideas to others. Communication of innovations is always a slow process in a traditional society. Where newspapers and others forms of mass media are relatively less important, oral communication remains the only way of spreading new ideas. In some villages of the Philippines, health workers have used village blackboards for spreading information and messages, but in the majority of rural communities oral communication is still the most important one. In this respect the village health worker is often an agent of social change because of the new ideas he/she introduces into the village.

Schools and school children can be developed into an important community resource. Each school child is a representative of a family and it is possible to teach parents and families through children. Furthermore, many school-age children participate in caring for their younger siblings at home and through them the care of the toddler can be improved. School children also participate in family activities on the farm or cattle grazing, protection of food crops from birds, harvesting and many others. Through them simple techniques like grain storage, soil conservation, composting, making soak-pits and so on can be introduced to families. Older school children can also become the nucleus of youth clubs, and instructions in good parenting can be organised for them. Finally, school buildings are important community institutions and can be utilised as focal points for community gatherings for seminars and study days. On such occasions the school vegetable garden can be used as a demonstration area.

The training institutions for medical auxiliaries and para-medicals are also usually to be found in the Districts. The students from these institutions can be utilised for small-scale community surveys which will be also valuable training for them.

People's skills are wasted if there is no channelling through careful community organisation. Formation of viable social institutions to support health programmes is often more important than putting up buildings for services. Here existing community groups can be approached. For example, in several countries of Africa the women's organisations are well developed and very influential. In countries with a one-party system of politics, the national political party has a great deal of social standing and is often active in social organisation. Local branches of such national organisations can be important allies in evolving programmes of health improvement.

Resources of material and labour

Many rural communities have traditions of mutual help. For example, in Indonesia such a tradition is well founded and often the village community gets together to help one of them with building, putting up an extension to a house, indoor work or other such activities. These traditions have been made use of by political leaders to promote national self-help schemes. Thus, the Harambe self-help schemes in Kenya have been responsible for setting up many local structures including large polytechnics. A very large proportion of Health Centres and sub-centres in Tanzania have been built through self-help schemes. More recently, in Jamkhed, India, self-help schemes have led to the construction of more than 20 small dams for water conservation schemes.

Besides resources of labour there may be other resources of material like wood (from local forests) and other products – stone (from local quarries), bricks (from local kilns) and so on which can be utilised for local activities, or as the focal point for setting up co-operatives and credit-unions. Such small-scale economic institutions will provide further support for community organisation.

Amongst local resources special mention should be made of food resources. Since a large part of human productivity in rural areas is concerned with food production, it is as important to rural life as money is to a cash economy. With the present drive for cash crops the raising of food crops may suffer. Hence, conservation of the local food resource in the form of a proportion of land earmarked by every farmer for growing the family's food supply is necessary. Many countries have utilised the attraction of cash crops to create communal farms for growing these crops (while food is grown on the family plot) and to form co-operative unions.

With regard to growing food crops, the farmer needs to know what proportion of his land to allocate to the growing of staple food crops and how much to the growing of other foods to supplement these staple crops. In the past undue importance has been placed on protein foods, including animal protein. No doubt protein has an important role in the body's economy, but with increasing understanding of the role of energy in the diet, the raising of energy-rich foods will be an important step in the creation of local food sufficiency. Ground nuts and soya are important sources of energy as well as protein. Coconut, oil palm, sesame, mustard and other sources of edible oil like cotton-seed should be encouraged if the soil and climate are suitable. In this respect oversight can lead to a shortage of essential foods with a consequent drain of resources.

Whilst considering the preservation and further development of food resources it is important also to think of human milk as an important resource in child nutrition. Studies in many parts of the world have shown that even the average undernourished mother is capable of producing between 400–600 ml of milk in the second year of lactation. This would amount to an average of 300 mg IgA and 10 mg IgG per day in addition to the protein, energy and other nutrients. This important resource is being rapidly eroded under high pressure advertising of baby food manufacturers and needs to be conserved through control of promotion and by raising community awareness. Countries like Papua New Guinea, Burma, Algeria and Guinea-Bissau have passed legislation aimed at conservation of this important resource.

Financial resources

Financial resources in health are inadequate everywhere but particularly so in developing countries where the average per capita health expenditure is one US dollar annually. This low level of government expenditure on health is likely to continue in the foreseeable future. Therefore, for developing District health programmes, the main financial resource will have to come from rational redistribution within the health budget. Already disproportionately large amounts of money are being spent on large teaching hospitals so that in some countries the recurrent annual expenditure of the teaching hospital is equal to that of the total health budget of the country. In the regions, the expenditure on curative care, mainly through the Regional and District Hospitals is again unacceptably high, leaving very little resource for development of rural preventive services. It is the general rule with most services that once a pattern is established, any withdrawal of a service, however extravagant, leads to an outcry. Reallocation of funds for preventive/promotive programmes which can lead to curtailment of curative services is likely to be difficult unless backed by strong political decisions.

Under the circumstances described above, many health workers have looked for alternative ways of financing rural health services. These approaches have their advantages in that they make the village health services independent of the competing demands on the health budget. They also ensure a continuing high level of health awareness in the rural population to be prepared to contribute for their services. And people are more likely to accept without reservations and utilise those services which they have created themselves. On the other hand, methods of local financing do raise the moral issue that irrational use of curative services slanted in favour of urban élites will be allowed to continue and the rural poor must find their own resources for health care!

Many health workers have come up with innovative ideas for generating local financial resources. Credit unions and health insurance have been successful in Indonesia. The former consists mainly of a revolving fund concerned with productive activities like agriculture or cash crops. An agreed proportion of the fund is earmarked for health. In the case of the scheme of health insurance, households contribute a fixed amount annually towards health and in return receive free preventive services and subsidised curative care during illness. In the Sudan, when capital was needed for a new health activity it was raised by imposing a small tax on long distance bus tickets. More than enough capital was raised in a few years to help several other community programmes. In all countries where decentralisation has been the government policy, development programmes are decided at the periphery and funds are allocated accordingly. This does not necessarily mean that wise decisions will be made, because the disparity between curative and preventive care will continue unless fundamental policy changes occur. In many cases decentralisation only means a shift from projects in the neighbourhood of the capital city to projects in the neighbourhood of the Regional and District centres. However, it does also ensure that the District Medical Officer will have a say in the decision-making process. The success of his Community Health Programme, his ability to establish a dialogue with the people and to create a climate of thinking as well as his leadership qualities will help in the allocation of funds for relevant health activities.

Private sources like charitable institutions, local plantations or industries, including mines, and co-operative dairies may provide financial support for local health activities. Very often their contribution is limited to their employees and sometimes to the families of the employees.

Considerable savings in finance can often be made. In a one per cent sample of outpatient cards, prescribing patterns at one health centre in a District were analysed and the average cost was calculated. These actual costs were compared with a standardised regimen and large savings were possible (see table 2.15).

It was estimated that 'appropriate' prescribing could lead to a saving of 70 per cent of the drug bill.

Natural resources

The productive life and economic activity of any settlement or community is based on existing natural resources. Based on these resources, and the skills to exploit the resources, a whole pattern of life styles emerges with one activity dependent upon another. Thus, the presence of minerals may attract mining industry with its own technicians, managers and clerks upon whom in turn the local businessmen and farmers depend. These natural resources are therefore important for the survival of the community.

The most basic resources are land and water on which agriculture depends. Conservation of the community's resources with adequate and rational utilisation of land will help increase productivity. Similarly, water resources can be conserved and improved in a variety of ways. In most countries of Africa there is enough land in rural areas and land hunger does not occur as in several countries of Asia and Latin America. In Asia, up to 40 per cent of a rural population may consist of landless labourers dependent for their livelihood on the vagaries of nature and the whims of the landlord. This situation is even more desparate in some countries of Latin America where less

Table 2.15 **'Actual' and 'appropriate' spending in primary health care**

	Symptom as % of a 1% sample	Average prescription costs 'Actual'	'Appropriate'	Possible % saving
Malaria	51	0.57	0.23	60
Cough	17	1.35	0.12	92
Measles	4	1.38	0.68	51
Diarrhoea	3	0.79	0.54	32

Source: IDS Health Group (1978). *Health needs and health services in rural Ghana*. Institute of Development Studies, University of Sussex, UK.

than 1 per cent of landowners occupy 42 per cent of the cultivable land. Most of these large tracts of land are used as ranches and plantations and the smallholders, who are the real producers of food, may have very little land available.

Equity in land ownership and land reforms may be beyond the scope of the District Medical Officer. But establishment of small plots of communal land for demonstration and teaching agricultural skills, and the use of the produce for pre-school feeding and for local nurseries, is an important part of nutrition rehabilitation. Several programmes of community health have found it necessary to branch out into the training of village agriculture workers including simple veterinary skills. These activities not only help to generate a new resource in the form of training for the peasant farmer, but also sufficient produce to support communal feeding programmes. Community awareness of local resources also leads to care of the environment chiefly in the planting of trees and the prevention of deforestation. The quality of the community's life is closely linked to the quality of the environmental and natural resources. Their preservation can often be made part of the community's responsibility.

Methods for finding out what is happening in the district

There are five main ways of finding out about communities within a Health District:

(1) The first method is to make a checklist of information required for reasonable planning and delivery of health services bearing in mind the human, material and financial resources that will be needed. The information can be extracted from available health statistics and from knowledgeable persons both within and outside the community.

(2) The second method is the use of existing data and records of health and health-related institutions in the District.

(3) Another method is the use of household surveys to obtain the socio-economic characteristics of the community such as housing, income, major economic activities and sources of food and nutrition which may have a bearing on health. During household surveys, knowledge, attitudes and practices relating to health problems should be obtained to provide the basis for health education plans, programmes and activities.

(4) Very occasionally, clinical surveys may be useful to obtain information on the health status of the community. Although such clinical surveys are expensive, they may offer the only alternative method of knowing about the prevalence and specific symptoms of important disease conditions in the community. Essentially, clinical surveys represent mass-screening procedures and as in mass-screening procedures, steps must be taken to make some treatment available for common illnesses which may be brought to light.

(5) Mass physical examination for assessing the health status of a community has been used in community health research projects. Useful as mass physical examinations in small communities may be, they can be very expensive. On the whole, mass screening procedures tend to raise the expectations of people examined and one must be ready to treat cases detected immediately if people are not to lose interest in continued participation in such community health action programmes.

The community round

A simple but effective way of finding out, at a glance, the likely health needs of a community is to do what is appropriately referred to as a community round. A community round can provide the same assistance to a community health worker as the clinician gets when he goes round the ward to assess the health needs of his patients and to prescribe treatment. Dialogue and interviews with key rural persons are useful in identifying local needs. The natural and elected leaders in the community are often used as key informants and it is advisable whenever possible to work through them. However, it has often been said that many of them represent one point of view and may speak with vested interests. In every society there also exist people who refuse to make use of the services, who disagree with the generally held opinions and are considered 'different'. Obtaining their views may also be of interest. Another group of key informants are the teachers. Teachers are keen observers of social and political currents in the community. Through their pupils they are in close contact with families and can often provide a balanced view of the community's needs as well as opinions about services. In every social group there are also individuals who are innovators and spread new ideas. It is important to identify such individuals and establish a regular dialogue with them. Also, as a method of obtaining reliable information 'group interviews' can be very helpful. When a person speaks in the presence of other community members, the information is likely to be more relevant and truthful on account of social pressure. The antenatal and under-fives' clinics provide many opportunities for such group interviews. Farmers' clubs, youth services and clubs, women's organisations and various religious groups are further examples of available sources for group interviews.

Why surveys may be required

There are three particular reasons why all illness is not necessarily reported to the health services. Many illnesses thought to be caused by evil spirits, breaking of taboos, or witchcraft will obviously be attended by the indigenous practitioners. Thus, in Uganda, kwashiorkor (obowesi) is thought to be due to jealousy between the child in the womb and that on the breast. Western medicine obviously has no cure for it and many cases of kwashiorkor are not brought to the notice of the medical services. In India, measles and several childhood exanthems are considered to be caused by the visitations of a deity and are not brought to the notice of the medical profession.

The second important reservation about health service morbidity data relates to the inverse care law mentioned earlier. Those in a community who need services most, utilise them the least. This is because they live too far away, or they cannot afford the fees, or they belong to a low social class, or they feel excluded, or have a major family discord causing unhappiness and depression, or the drudgery of earning a living leaves them with very little free time and energy to do anything else. Hence a special effort needs to be made to identify such groups and their most prevalent health problems in order to obtain a more complete picture of the pattern of ill health in the District.

Thirdly, in many countries where most diseases are acute, the morbidity data of health institutions mainly reveal attendances for such illnesses as was shown in table 2.10 and figure 2.9. It seems that chronic illness is not considered important enough by the people and the health profession alike to merit a great deal of time and attention. People may have learnt that little is done if they bring such problems to the health units. Thus, cases of established polio paralysis, cerebral palsy, blindness, and other handicapping or chronic illnesses are rarely brought to the health centres or hospitals for treatment. Visits to villages and surveys of pre-school and school children are essential to obtain information about the prevalence of such health problems (table 2.16).

The above reservations should be borne in mind when planning surveys. A key factor in their success is generating good questions, which requires the identification of key issues in a topic. An example of key issues in child feeding is given below.

Finding out about child (0–5 years) feeding

How do people feed their young children?
    Breastfeeding? Food and meals at different ages?
    Quantity offered and eaten? Energy density?
    Nutrient values? Frequency of feeding?

Are child-feeding practices 'satisfactory'?
    Identify the areas where problems are known to occur.

Why do people feed young children in the way they do?
    What do we see if we spend three days in each season
    observing and participating in local family life?

How much trouble would it be for people to change? What would be the cost?

What do people see as their main problems in feeding young children?

If people do change their behaviour, what will the consequence be? Will it work in their situation?
    What happens when mothers use a new weaning food and find their children have diarrhoea?

District health needs

Table 2.16 Surveys – are they useful or not?

1 Usefulness of the study depends on applicability and cost effectiveness.
2 Data are often available. Is utilisation of scarce resources, time and manpower justified?
3 Will the information obtained help with:
 • planning and implementation of agreed health strategies?
 • monitoring of programmes and corrective action when needed?
 • evaluation of programmes?
 • assessing health needs with a view to making managerial decisions rather than clinical information?
4 Are the questions being asked valid and clear?
5 Will expectations be raised which cannot be fulfilled?

What do people think they will get out of changing?
 Prestige? Convenience?
 Healthy and attractive children?
 More productive children?

How are people changing their child-feeding practices already?
 What happens when grandmothers and daughters talk about the way each has fed (or is feeding) their children?

Another example of the use of surveys to assess prevalence is in the case of residual paralysis after polio. A postal enquiry to school teachers in areas where at least 70 per cent of children attend school has been shown to be an efficient method of measuring the prevalence of polio. This method first developed in Ghana has now been tested in several countries.

Surveys of health service delivery problems – operational research

In the actual process of delivering health care, it is important that procedures be constantly monitored to provide information for the reappraisal of desired goals so as to prevent programmes from being shipwrecked.

Operational research aims at applying management techniques for the improvement of health care delivery. In a way it is a method of on-site evaluation for staff engaged in the actual process of delivering health care without waiting for outside agents to detect faults after irreparable damage has been done.

One effective way of conducting operational research in clinics, Health Centres, health posts and other District and sub-district health institutions is the use of trained observers to administer appropriately designed questionnaires or checklists while observing processes resulting from the interaction between health providers and patients as well as other persons attending special sessions at health facilities.

When direct observation procedures as described above have been refined and standardised, it should be possible for health facility staff to conduct a retrospective audit of clinic records and institute immediate action to improve the running of their health facility.

In summary, the pattern of epidemic, acute and life-threatening disease can be obtained from hospital and clinic statistics. Information on the age distribution of the population can be obtained from local enquiry and confirmed from census data. When information is needed about the pattern of other prevalent illnesses, particularly chronic and handicapping conditions in rural areas where there are no clinics to provide even approximate statistics, a supplementary method for obtaining information is required. If the sampling technique of the survey is well designed, a small number of subjects will provide information representative of the whole group or area. Information on morbidity thus gathered can be further added to and improved by means of interviews with community leaders, other residents and practitioners. Surveys of practices and attitudes may throw further light on beliefs about disease. Naturally, this source of information is not available immediately. As gradually trusts and relationships get established, and friendships are built, more reliable information of this type can be gathered.

Resources can be identified by mapping existing health care facilities and staff. Enquiry will add the whereabouts of other practitioners in the District. Attendance can be examined to find out where people come from. Gradually information can be assembled on other resources affecting health and health care; people (in agriculture, education, and so on); equipment and time spent; natural resources, and so on.

FURTHER READING

Ebrahim G. J. *Social and Community Paediatrics in Developing Countries*. Macmillan Press Ltd, London, 1985.

Enkin M., Kierse M. J. N. C., Chalmers I. *A Guide to Effective Care in Pregnancy and Childbirth*. Oxford University Press, Oxford, 1989.

IDS Health Group. Health needs and health services in rural Ghana. *Soc. Sci. Med.* (1981) **15A**: 397–495.

Kleczkowski B. M., Elling R. H., Smith D. L. *Health System Support for Primary Health Care*. WHO, Geneva, 1984.

Lutz W., Chalmers J., Hepburn W., Lockerbie L. *Health and Community Survey*, Volumes I and II. Macmillan Press Ltd, London, 1992.

Selwyn B. J., Frerichs R. R., Smith G. S., Olson J. (Eds). Rapid epidemiologic assessment. *Int. J. Epid.* (1989) **18**: Supplement 2.

Vaughn J. P., Morrow R. H. *Manual of Epidemiology for District Health Management*. WHO, Geneva, 1989.

Werner D., Bower B. *Helping Health Workers Learn*. Hesperian Foundation. P.O. Box 1692, Palo Alto CA.94302 USA. 1982.

World Health Organization. *Uses of Epidemiology by Front-line Workers*. WHO, Geneva, 1981.

3 Making a Health Plan for the District

With the trend towards decentralisation, the Districts are experiencing a certain degree of autonomy. Under the previous system of tight central control the health systems of the District had very little scope to respond immediately to local needs and adapt to local situations. The decision makers were too distant from the front-line. As a result of decentralisation, new lines of division of responsibilities are being drawn between the centre, the regions and the Districts. Obviously, these will differ from country to country depending upon the political structure and the historical past. But in general, the areas of responsibilities are as follows:

Ministries of Health are chiefly responsible for:

(1) Health policy formulation, including type and extent of intersectoral collaboration.
(2) Production of national health plans and guidelines for regional and local planning.
(3) Allocation of resources, particularly capital funds.
(4) Provision of specialist advice for specific programmes.
(5) Purchasing of pharmaceutical and medical supplies for the country as a whole.
(6) Control over health manpower development.
(7) Regulation of the private sector and NGOs in support of the national policy of PHC.
(8) Control over national research institutes and other health organisations.
(9) Liaison with international agencies.

Provinces and regions carry responsibility for:

(1) Regional level health planning in accordance with national policy guidelines.
(2) Co-ordination of all regional health activities.
(3) Provision of supplies and other logistical support.

(4) Managerial and technical supervision of all health manpower in the region.
(5) Collaboration between health and other sectors in the region.
(6) Budgeting and auditing of health expenditure in the region.
(7) Obtaining from the Ministry of Health approval and financing of all large capital projects.

Districts are assuming the following responsibilities:
(1) Preparing and implementing a health plan for the District.
(2) Organisation and running of all public sector health facilities in the District.
(3) Co-ordinating the work of the private sector and NGOs in line with the overall health objectives for the District.
(4) Implementation of all community based health programmes.
(5) Management and control over local health expenditures.
(6) Promotion of active links with other sectors in the government.
(7) Promotion of community participation in local health activities.
(8) Collection and compilation of local health information for forwarding to the regional office.
(9) In-service training of health staff.
(10) Training and supervision of all community health workers.

As Districts have assumed greater autonomy, a number of key weaknesses in the District health systems have come to light. These are:

(1) Organisation, planning and management expertise is commonly weak in the District.
(2) Goals, targets and procedures are poorly defined.
(3) A great deal of information is collected routinely but not analysed. At the same time, information required for decision-making and monitoring is not systematically collected.
(4) Vertical programmes do not get integrated into an overall District Health Plan.
(5) Evaluation of programmes and health systems are often carried out by consultants, academics or headquarter staff without involvement of District Health Managers. As a result the studies are not appropriately oriented and results not practically applicable.
(6) There is very little action research.
(7) Communication between health facilities and programmes and with local communities remain weak.
(8) The orientation of health workers continues to be clinical.
(9) Logistical support is erratic and weak. Transportation which is the key to support and supervision is poorly managed.

In developing a robust District health system these weaknesses listed above, and many others, will have to be addressed.

The five pillars of the District health system are:
(1) Planning, organisation and management (discussed below).
(2) Financing and resource allocation to achieve plan targets (discussed on page 114)
(3) Intersectoral action (discussed on page 105).
(4) Community involvement (discussed on page 102).
(5) Development of human resources (discussed on page 100).

Planning organisation and management covers several activities like programme planning within the overall national guide-lines; agreeing goals and responsibilities of different organisations and health teams in the District; co-ordination of programmes and activities within the health and other sectors of the government as well as private, non-governmental and community organisations; planning of individual programmes; provision of supplies, equipment, drugs and transport for the efficient running of programme activities; and manpower planning.

WHAT IS A PLAN?

A plan is a course of action one intends to follow in order to achieve defined goals and objectives. It ensures that targets are specified and best use is made of available resources. A plan also provides the opportunity to consider all the options available for the performance of tasks in response to needs. In other words to plan is to choose from a number of available options in order to decide upon a course of action for achieving desired objectives. Thus, the word planning denotes a step-by-step process in achieving a purpose. A District Health Plan provides the rationale for choices about deploying and utilising available resources to achieve short-term and long-term objectives.

Thus planning is about the *future*, described not in vague terminology, but as well-defined *goals*, and with a *time-scale*. This calls for an innovative and imaginative approach. But planning also takes a hard look at reality by taking into account the *available resources* and their use in the most effective and efficient manner. A plan may operate over several years. In order to make sure that the current of events does not make one drift too far away from the desired objective there has to be a process of monitoring and continually reviewing the planned activities.

The above description of the planning process embodies several key words and phrases which need to be seen in the context of the District health services, and a brief description follows.

Future

Future implies a long-term view. Considering staff movements and rapid socio-political changes which are common in developing countries, it is more practical to think in terms of five year plans. The plan should be in operation long enough for key staff members to be able to come to grips with it, and yet robust enough to continue when key staff are transferred to another post. For the foreseeable future a District Health Plan will largely relate to the different elements of Primary Health Care; or the control of specific diseases for example diarrhoea, tuberculosis, leprosy and so on; or it may relate to coverage and quality of services. In a broadly based plan it is best to think of the key items separately as a series of mini-plans which can then be amalgamated into a unified health plan. Under each topic one begins with identifying what one wishes to see happening in five years' time (the objective), stating it in realistic terms for example 'increase in availability of clean water by 20 per cent *or* 'raising immunisation coverage to 80 per cent' *or* 'set up prenatal services in remote communities' and so on. Having set the five year goal in the activity of choice, one then works backwards in the same manner and states what should be happening in three years' time as a step towards the five year goal. Such a step-wise method of planning helps to set the time scale for example short-term (three year targets), and long-term (five year target). In this way there is a sequencing of activities which sets a pattern for the deployment of resources.

It helps to bear in mind that in all plans that deal with populations one is always trying to strike a moving target. Populations change because of births and deaths, in- and out- migration, and so on. Allowance should be made for these changes so that the plan does not stray too far from the desired goals. Similarly, conditions like drought, heavy rain and floods or famine can pose emergencies requiring urgent divergence of resources. Inflation is a major problem in most developing countries and causes major shortfalls in funding. Hence it is necessary to keep the plan under constant review.

Goals

A plan is intended for achieving defined goals. A definition of goals at the outset is the first step in good planning. Goals should be stated in clearly measurable terms, for example, 'increasing antenatal coverage rates to 90 per cent from the current 50 per cent'. A general statement like 'further improvements in antenatal coverage' is not enough. There may be several intermediate and subsidiary goals, and all should be stated in the same manner.

Having set the short-term and long-term goals in realistic and measurable terms the next step is to outline the *activities* needed to achieve the goals, the *resources* required for supporting the activities, and how the resources will be obtained. As activities commence, it will be necessary to have indicators, and milestones that will demonstrate what progress is being made.

At the planning and implementation stage there are choices to be made between different approaches and strategies. Resources are always limited, and rarely arrive on time. Alternative strategies are then called for in order to continue to maintain momentum. In such a case choice has to be made between strategies. When confronted with such a situation the choice is based on what is most important, what is urgent, the cost, and ease of achievement as balanced against community concern, and its effectiveness as a health measure. Each choice has advantages as well as disadvantages, and these should be carefully considered.

Planning may be thought of as a process both at the central and at the peripheral level. At the central or policy-formulation level, a plan considers the broad outline of what needs to be done and the resources that must be committed at the National level to enable policy objectives to be achieved. On the other hand, at the peripheral level, details of actual tasks to be carried out, and who is to perform them, must be considered in the face of real everyday problems of resources and organisational difficulties. Planning at the periphery, namely at the District or community level, needs to be more pragmatic and less conceptual than at the National or the Regional level.

Planning at the central and peripheral levels must not be considered as separate exercises because information obtained from the periphery is required for central planning. Similarly, national guidance and policy are also needed in peripheral planning. Planning based on needs identified at the periphery is sometimes referred to as the 'Bottom-Up' approach in planning.

By their very nature, national health plans tend to be based on aggregate information. They address general concerns and matters of policy without necessarily focusing on specific issues district by district. Moreover, since resources are limited, national targets must be set in a way that takes into account a fair distribution of available resources over the country as a whole. National Health Plans tend to describe issues like equity, accessibility, intersectoral action, community involvement, integration of vertical programmes, and so on in the most general terms. There is so much diversity from one part of the country to another that it is impossible to do otherwise. The District Health Plan has a greater scope for entering into details of issues and provides specific objectives, targets to be achieved, the time scale over which they are to be achieved, the strategies to be followed with details of budgeting, monitoring and evaluation, and so on. A District Health Plan which is logically developed, well written, and specific in content stands a better chance of support from the regional and national planners than the one describing only generalities.

THE PLANNING PROCESS

The planning process entails the development of consensus within the health team, and with other governmental departments whose activities are essential for developments in health as well as agencies and organisations operating in the District. Sharing of a vision of the future and setting up a process for translating this vision into concrete plans is an essential element of management. In order to achieve this the following actions are needed:

(1) Wide dissemination of information about national guidelines and priorities as well as of goals, objectives and strategies.
(2) A dialogue amongst the providers on the one hand, and between the providers and the community on the other regarding the main health problems.
(3) Formulation of plans based on the consensus and their general dissemination for feedback.

When a general plan is agreed by all the parties concerned, the next step is to develop:

(1) action plans which specify activities, targets and the time frame in which targets will be achieved, as well as assigning clear responsibilities to teams and individuals.
(2) the setting up of managerial systems supportive of the action plans (Chapter 4).

The health planning and implementation cycle

The essence of effective planning lies in finding answers to four key questions:

(1) Where are we now? (District assessment)
(2) Where do we want to go? (priorities, goals, targets, decisions)
(3) How will we get there? (organisation and management)
(4) How will we know when we arrive? (evaluation)

The steps in conducting a District assessment of health problems, resources and opportunities were discussed in chapter 2. Only when a searching assessment has been done can the next steps be taken. These are to establish health priorities; then to identify key tasks for actions, their organisation procedures and required inputs, and to put the plan into action with built-in feedback mechanisms (see figure 3.1).

Dangers of planning and why planning sometimes fails

A formal planning system can have dangers as well as advantages, and these need to be recognised and avoided. One possible problem of 'over-planning'

90   District Health Care

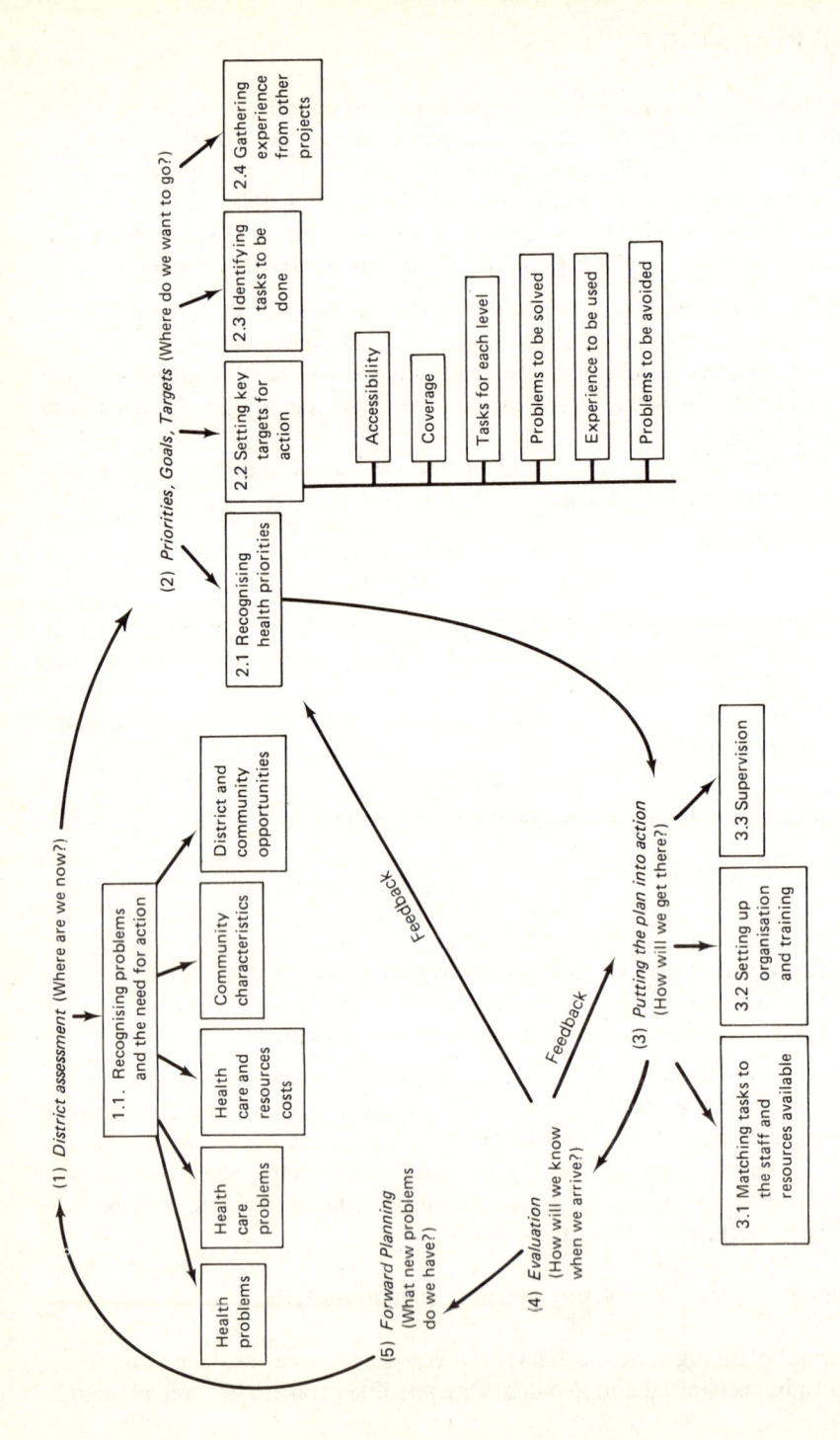

Figure 3.1   The health planning and implementation cycle

Table 3.1 **Some advantages and pitfalls of formal planning systems**

Possible advantages	Possible disadvantages
1 Action is called for; no action is immediately apparent.	1 'Over-planning' and inhibition of creativity and innovation, especially when front-line health workers and the community are excluded from the planning process.
2 More people selectively involved to do specific tasks.	
3 Many relevant factors considered.	
4 Common or traditional assumptions challenged and questioned.	2 Too many people or inappropriate people involved.
5 Unity of direction and purpose.	3 Wrong focus.
6 Encourages use of a range of possible strategies.	4 Takes up too much time.
	5 Insufficient time available to plan properly.
7 Plan can be evaluated.	6 Curtailment of flexibility.
	7 Strategies may be considered in the wrong sequence.
	8 Plan may be too broadly defined to be measurable.
	9 Inhibits action on immediate problems.

(Adapted from Camillus, J. C. (1975). *Evaluating the benefits of formal planning systems, long range planning.*)

may be that there is little scope for flexibility and innovation after the plan is written. Alternative approaches for solving a problem may be forgotten in the bustle of making a decision to do something. Good ideas may have no mechanism for being brought forward and put into action. People may feel there is little scope for innovation and creativity locally. Another problem is that too many people may be involved. Alternatively, inappropriate people may be involved. This is particularly likely when planning is over-centralised and done without the participation of local people or the staff at the periphery who will be responsible for putting the plan into action. Another problem can be that an elegant plan can be completely misconceived and wrongly focused. The planning process can also absorb too much time so that most of the time available is taken with drawing up the plan and there is no time left to tackle the important issues associated with putting the plan into action. When this happens it is easy to get strategies in the wrong sequence and if there is no flexibility in the plan this will cause considerable difficulties. Even though a number of problems can be envisaged with a forward planning system there are still many advantages in developing a planning process. These are compared with the pitfalls in table 3.1.

Plans may sometimes fail. It is by understanding why such failure occurs that planning processes and plans themselves can be improved. Table 3.2 lists some of the reasons why planning may fail. The three main reasons for failure

Table 3.2 **Why planning sometimes fails and what can be done**

Why does planning fail?	*What can be done?*
1 Lack of commitment by key managers	
1.1. Lack of acceptance by 'operational' personnel who will put the plan into action.	Involve 'operations' people in drawing up plans.
1.2. Lack of interest and commitment by senior personnel.	Ensure commitment before starting to plan.
1.3. Some managers are allowed to opt out.	Ensure involvement of all managers at an early stage.
1.4. Confusion about what 'corporate' planning means.	Get commitment to the District health activities by all segments involved in health and get all these views represented in the plans made.
1.5. Planners ignore the work people are already doing.	Find out what people do and the problems they have before recommending any change.
2 Poor planning processes	
2.1. Plans are made centrally in 'ivory towers'.	Hold discussions with people working at the periphery.
2.2. Confusion between strategic and operational planning.	Distinguish broad policy making and strategy from the details of putting a policy into action.
2.3. Trying to plan through committees.	Set up a working group instead.
2.4. The planner is of too low a calibre.	For negotiation find someone who is acceptable to most of the segments that will be involved in putting the plan into action. For the hard routine of working out the plan's implications find someone who will do the job well – they need not be the negotiator.
2.5. The planning system and the plan is too sophisticated and too complex.	Make it simple so it fits in with people's current work.
2.6. Planners fail to accept the limitations of their role.	Distinguish between planning and the 'operational' activities of putting plans into action.
2.7. Insufficient attention is given to the format of plans.	Use a simple format and seek the views of others.
3 Plans are not used	Implementation needs to be built into the plan. Commitment is needed before plans are drawn up, and maintained during implementation.

(Adapted from Hussey, D. (1974). *Corporate planning: theory and practice.* Pergamon Press.)

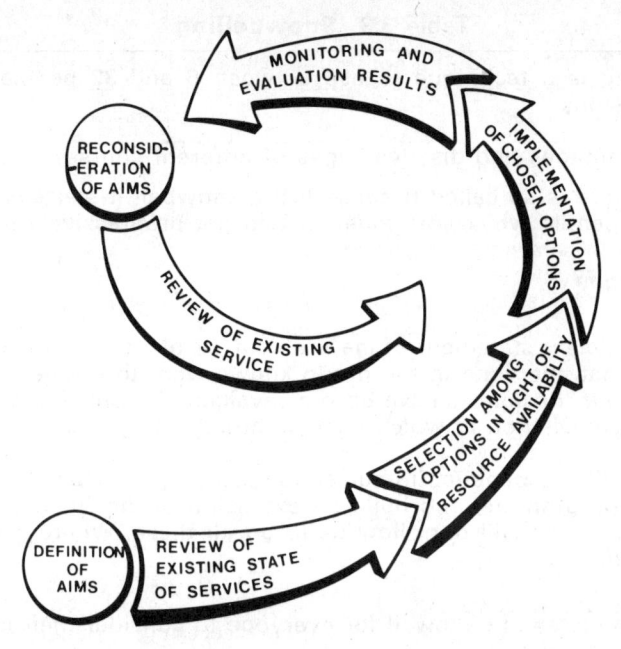

Figure 3.2   Planning is a learning process

are: (i) lack of commitment by key managers, (ii) poor planning processes, and (iii) plans remain on paper and are not implemented.

Planning is a learning process

Planning is a continuous learning process. Aims need to be constantly reviewed as the pattern of health problems changes, as opportunities arise for more effective use of resources, and as the focus of priorities and policy moves. This continuous process is illustrated in figure 3.2.

Key concepts in effective district health planning

(i) *Priority health problems determine health service functions*

Once the priority health problems are recognised, and the conditions, causes and risk factors contributing to them identified, it is possible to define clearly what the functions of the health services need to be. As described in chapter 2, the priority health problems are selected on the basis of their prevalence, seriousness, preventability and treatability. Another technique for deciding priorities is 'snowballing' (see table 3.3). Table 3.4 shows an example of priorities in planning for child health services. Naturally, when the health problems change, a new plan with new health service functions is needed.

Table 3.3 **Snowballing**

Snowballing is a technique to use between 8 and 32 people to help reach decisions
or
clarify advantages and disadvantages of different courses of action.

The technique is so called because like a snowball it starts with small groups of people which join together and get progressively larger.

The technique

Stage 1
Start with a clear statement of the problem, or of alternative courses of action so that everyone in the group knows what to discuss.
An example is 'Resources have become available for building one Health Centre in the District. In which village should it be built?'
or
'We would like to conduct immunisation campaigns, expand the health education programme, and build an extension of the District Hospital. But the resources will not allow us to do all these. Where should our priorities lie?'

Stage 2
About 10 minutes are allowed for everyone to consider their ideas and note down the reasons.

0 0 0 0 0 0

Stage 3
Individuals form groups of two. They explain their points of view to each other, and then *must agree a joint point of view*.

00 00 00 00 00

Stage 4
The pairs now join together to form groups of four people. The pairs present their joint view in turn, *and groups of four must reach a joint agreement*.

0000 0000 0000

Stage 5
The joining together process continues and larger groups are formed. The groups of four present their decisions and *the group of eight reaches a joint agreement*.

00000000

Stage 6
The groups of eight present their decision to the rest of the meeting in a plenary session.

Why does snowballing work?
The strength of the technique lies in *participation*. Everyone has to formulate a view and express it at least once. Also everyone has to listen to another point of view and find a common ground. The atmosphere is also less threatening.

Table 3.4 **Priority health problems and functions of health system: child health**

Priority health problems		Basis for selection	Functions of child health services
			Supervise and maintain health of young children by:
Group 1:	Malnutrition		
	Malaria	⎫ different mixtures of	(1) *Promotion of nutrition* – nutrition education on food needs and feeding practices (primary prevention of malnutrition)
	Severe chest infections, mainly pneumonia	⎬ prevalence, seriousness, preventability and treatability	– supervision of growth and early detection of malnutrition (secondary prevention)
		⎭	– supplementary feeding when necessary
	Measles	⎫ prevalent, serious, preventable and/or treatable	(2) *Primary prevention of infectious diseases through:*
	Acute diarrhoea	⎭	– immunisation
Group 2:	Neonatal tetanus	⎫ serious	– prophylactic drugs (e.g. antimalarials, treatment of contacts)
	Polio	⎬ fairly prevalent but	– health education
	Tuberculosis	⎭ easily prevented	
Group 3:	Intestinal parasites	⎫ prevalent, debilitating	(3) *Management of common childhood infections (as above list) by:*
	Skin infections – wounds	⎬ preventable and/or treatable	– early diagnosis and effective therapy (including health education on early symptoms and home treatment)
	Conjunctivitis	⎭	

Source: IDS Health Group (1978). *Health needs and health services in rural Ghana.* Institute of Development Studies, Brighton, UK.

(ii) *Health service functions determine what accessibility to services is needed*
If health services are to provide care effectively, certain services (such as monitoring children's growth) need to be available very close to every home in the local community. Other services (such as blood pressure measurement during pregnancy) need to be within 4–5 miles (8 km) or two hours travel, possibly at a small health station; and for certain emergencies as well as for some specialised types of care, a District Hospital needs to be accessible, ideally not more than three hours travelling time away (see figure 3.3 and table 3.6).

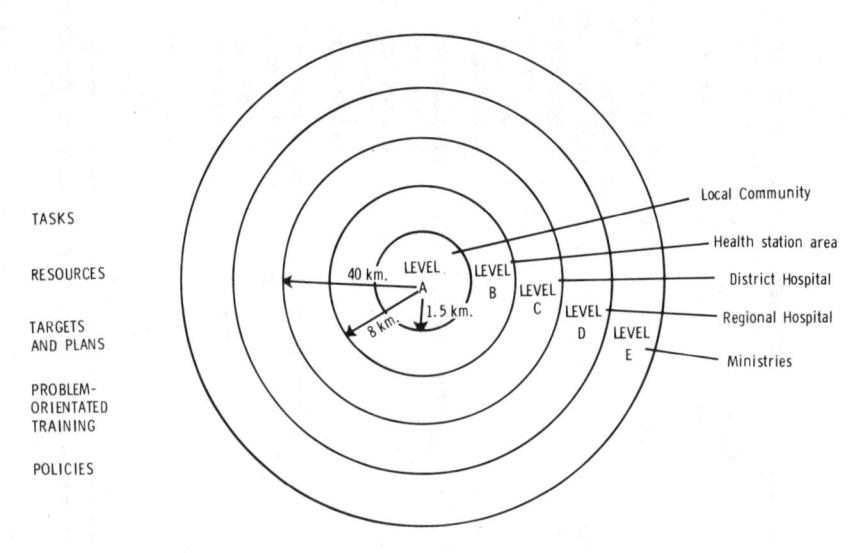

Figure 3.3   The over-all health care system with health functions determining accessibility to services

(iii) *Health service functions also determine task definition and evaluation*

Once health service functions based on priority problems are recognised and the accessibility of various services determined for each of these functions, tasks can be defined as components of such functions (see table 3.5). These tasks can then be allocated where services are needed and with the resources available. Table 3.6 gives an example of task specification for the control of malaria, measles and tuberculosis. It also shows the division of these tasks into three levels, the Local Community, the Health Station, and the District. This task definition and allocation with the resources available is in itself a tool for evaluation of what is being done. Moreover, such an allocation of tasks takes into account the fact that the control of malaria requires a number of health activities by different departments, including, for example, environmental sanitation, health education, curative services, and so on.

District health plan

(iv) *Health problems require action on causes of ill health, early intervention and rehabilitation as well as cure*

Most health problems require action at several stages:

(1) *Health promotion* – to maintain healthy life-style.
(2) *Action on causes* of ill health before a health problem begins (for example, improvement of a household's eating patterns, clearing mosquito breeding sites, removing health hazards, providing immunisation).

Table 3.5 **Principles of task analysis for planning health care**

---

1 Identify *priority health problems and the functions* for health services implied by them.
2 Use the concept of *primary, secondary and tertiary intervention* to identify the tasks to be done.
   Primary prevention: Action on cause of ill health
                         Promotion of health
                         Community development

   Secondary prevention: Screening for early disease
                         e.g. growth faltering in children
                         unsymptomatic hyperglycaemia
   Tertiary prevention: Early treatment
                         Also rehabilitation when necessary
3 *Specify all tasks* needed to tackle particular problems in the District e.g. malaria, diarrhoea, respiratory infections etc.
4 *Allocate the tasks to the resources available* (e.g. the local community (Level A), health station (Level B) and District level (Level C).
5 *Identify current resources* not being fully used for health care and health promotion (e.g. TBAs, herbalists, mothers, school children, community development personnel, etc.). Also *identify current problems in the health care system (especially at health centre level)* so these can be solved. Otherwise they get perpetuated into any new health care system in the local community.
6 Find out the *experience of setting up* new programmes (e.g. TBAs, CHWs) and find out *what can go wrong*.
7 Find out the *experience of training* people for new programmes *and* find out *what can go wrong*.
8 Set *priorities for training*, dividing tasks to be done into Level A, Level B and Level C training.
9 Calculate the work to be done in a specified population, local community, health station and the District.
10 *Timetable the tasks* to be done *each month*; calculate the hours of work and *calculate the number of people* needed to fulfil the tasks.
11 *Calculate* regional and national manpower requirements and plan sequence of training programmes.
12 *Think* of the implications if Level A knowledge and skills are to become part of the general knowledge of most people in 20 years' time. What is the implication for schools now?

Table 3.6 **What actions are needed in the district for communicable disease control**

	Malaria	Measles	Tuberculosis
LEVEL A The local community	1 Cleaning breeding sites Burying tins etc. Draining swampy areas and waste water Keeping eaves water butts covered etc. 2 Distribution of antimalarials (and keeping records in family register) to under fives to pregnant women 3 Treatment of acute attacks of malaria and recording each episode in family register Sponging for high fever Chloroquine tablets	1 Promotion of better nutrition 2 Case finding 3 Advice on home care treatment (especially feeding often) 4 Treatment and/or referral of complications depending on severity 5 Organisation for immunisation sessions for children over nine months 6 Records in family register of incidence	1 Keeping records of cases and contacts in family register 2 Following up cases to ensure continuous treatment – need to explain so that individuals understand the necessity 3 Education on coping with the disease (no spitting etc.) 4 Case filming of chronic cough 5 Contact tracing
LEVEL B The Health Station	1 Treatment of severe attacks of malaria referred from 'A' 2 Drug supplies for Level 'A' 3 Collect records from Level 'A' 4 Spraying campaigns in local communities	1 Immunisation (at family health sessions if possible) 2 Treatment of complications 3 Training level 'A' to recognise measles and to teach people measles care at home	1 Regular teaching and updating of Level 'A' tasks 2 Supplies distribution to Level 'A' (oral tablets) or treatment and record keeping (at family health sessions if possible)

5 Training for Level 'A' (see box above)

3 See cases occasionally (with DMO) to monitor:
   · weight gain
   · continuation of cough
   · sputum analysis
4 Sputum samples from chronic cough (at family health session) to be sent to District for analysis
5 Referral of suspicious and confirmed cases
6 ?BCG for newborn and under 15s

4 Collection of records from Level 'A'

LEVEL C
The District

1 Treatment of severe cases referred from 'B'
2 Monitoring incidence in District
3 Maintaining spray supplies
4 Maintaining drug supplies and teaching to 'B' and 'A'

1 Treatment of severe cases (expected referral needs to be estimated)
2 Monitoring incidence in District
3 Maintaining supplies and support for training 'A'
4 Possibly organising occasional District campaigns

1 Diagnosis of sputum samples
2 Start treatment
3 Information to Level 'B' for Level 'A' for following up contacts and continuity of care
4 Discharge of all patients
5 Support for Level 'B' to teach level 'A'

99

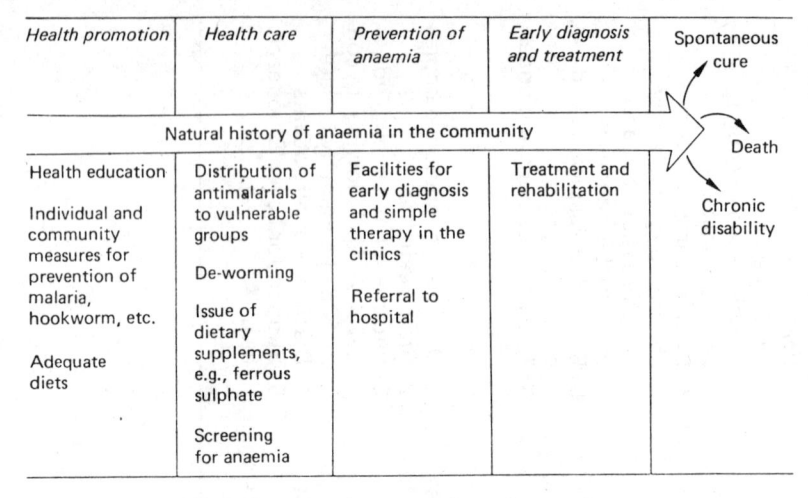

Health promotion	Health care	Prevention of anaemia	Early diagnosis and treatment	Spontaneous cure
Natural history of anaemia in the community				Death
Health education Individual and community measures for prevention of malaria, hookworm, etc. Adequate diets	Distribution of antimalarials to vulnerable groups De-worming Issue of dietary supplements, e.g., ferrous sulphate Screening for anaemia	Facilities for early diagnosis and simple therapy in the clinics Referral to hospital	Treatment and rehabilitation	Chronic disability

Figure 3.4 Interventions for the prevention of anaemia

(3) *Early intervention to prevent the problem becoming serious* (for example, health surveillance of children and pregnant women and early treatment of infections, recognising high risk mothers and children and families and providing special care).
(4) *Cure* if possible when the problem does arise (for example, proper care of the sick and injured).
(5) *Care and rehabilitation* after the problem has gone (for example, provision of aids for mobility for post-polio patients).

Figure 3.4 shows how the different types of interventions described above are intended to alter the prevalence and seriousness of a disease, in this case, anaemia.

(v) *Functions and task definition determine the pattern of staffing*

The aim in manpower planning is to get the right person in the right place at the right time at the right cost. By using task definition, a right person can be specified who will be able to do the needed job. When tasks have been specified, evaluation of what has been or has not been done is also made very much easier. Task definition and evaluation have in the past often been left out of health plans since they tend to fall in a no-man's-land between health planners and the trainers of health workers (see figure 3.5).

In the past staffing was often decided by the population: health worker ratios, for example, the number of doctors or nurses or auxiliaries per unit of population. There were three major problems with this population ratio

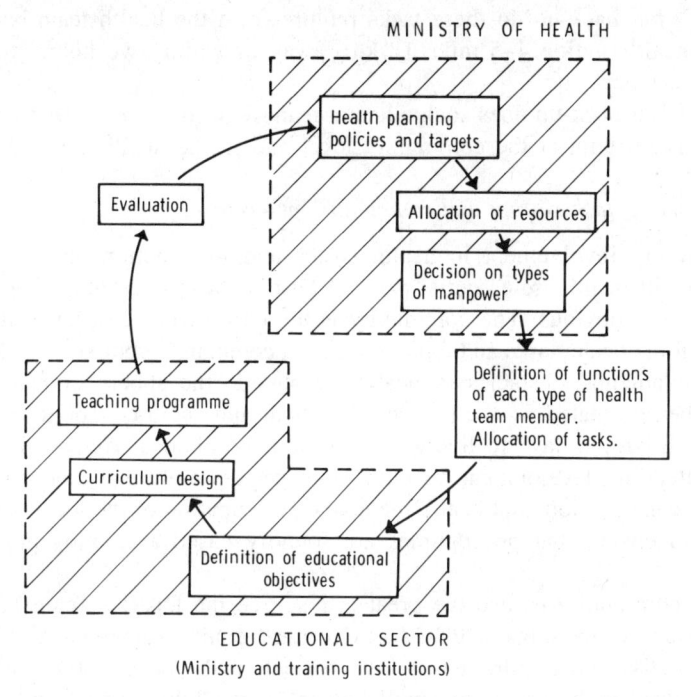

Figure 3.5   The usual no-man's-land of task definition and evaluation

approach. The first problem is that of inequitable distribution. Although the National or Regional or District ratio of doctors to population may appear reasonable, it may hide gross differences in different parts of the District. Secondly, such a simplistic approach fails to take account of the accessibility of the health workers to the population. Few people travel for more than two hours (or 4–5 miles, 8 km) to a Health Centre or a sub-centre. It is not realistic to assume that Health Centres and similar facilities provide effective care beyond a distance of two hours' travel on foot. Thirdly, a population ratio approach may assume that mere provision of health manpower can provide good health. The tasks that the health workers perform may not be taken into account and the logistic support essential for efficient functioning may also get overlooked. For these reasons the 'functional approach' to health manpower planning has evolved.

The new 'functional' approach to District staffing and health manpower planning has as its starting point the three linked questions:

(1) What health tasks need to be done and can be done within the household and in the immediate community within a distance of one mile (1–2 km) from every home?

(2) What back-up do these tasks require from the health team based at a health station 4–5 miles (8 km) away or within two hours' travelling distance?
(3) What back-up does such a health team require from the District Health Team both at the community level and in the health station?

(vi) *Linking task setting and community involvement*

Community involvement in health issues and joint action on the determinants of ill health are essential parts of District health planning. The reasons for this are obvious. The community is the core of the District health strategy. All the health tasks and functions have a community component. Moreover, the fundamental obstacles to health lie outside the sphere of medical science, being created by the physical, political and socio-economic conditions in which people have to live and earn their livelihoods. Across all the sectors affecting development like for instance, agriculture, water supply, resource conservation, and health the active participation of the local population helps to ensure that programmes are soundly based and enjoy public support.

The community is also the greatest resource the District Team has available. Many of the aims of PHC like coverage, equity, self-reliance, efficiency and effectiveness require active involvement of the people. By working together the health care system will become part of the community's responsibility. It will not be seen merely as something remote run by the Ministry of Health. Hence in drawing up a District Health Plan the challenge is to involve the local community in the process of task analysis, and in the process of deciding best approaches for getting certain tasks done. There is a greater possibility of projects and programmes to meet the needs defined by the people as compared to those defined by the health sector. Linking services to locally agreed health targets, and managing with the support of local people improves the chances of achieving the plan objectives.

In most countries, however, theory is running ahead of practice. This is because community participation has been interpreted in three different ways viz.

(1) *contribution* to projects and programmes;
(2) *organisation* of various social and administrative structures; and
(3) *empowerment* through knowledge and skills transfer for managing local activities more effectively, as well as for deciding and taking initiatives on local matters important to health.

In the past there has been over-emphasis on putting up structures. This has worked to a certain extent. However, it is more important to strike a balance between activities aimed at creating awareness, and those aimed at community mobilisation. The aim is not merely to mobilise people and resources for a given purpose, but also create a process whereby people can increase their control over the social, political, economic and environmental factors affecting their health status.

Bearing in mind that a community is a geographical, as well as a social and political entity comprising many interest groups, just having a Health Committee is not enough. Only the traditional or hereditary leaders or the more powerful get appointed to it. For canvassing opinion across the whole cross section of the community the representation of all sections must be sought. One should also go for a variety of informal social structures catering for different interest groups like, for example, the farmers (e.g. farmers' clubs); the women (e.g. women's groups); the youth (e.g. youth clubs), and so on. One problem is that it is often forgotten that in dealing with such groups the principles of adult education apply viz.

(1) Adults have a wealth of knowledge and experience waiting to be tapped and channelled.
(2) Adults learn better by doing.
(3) Adults have developed a strong self-image which they will protect rather than risk embarrassment in front of others.
(4) Adults grasp issues less quickly than younger people.
(5) In many communities large numbers of adults have been secluded from outside ideas, and there may be many barriers to cross.

In drawing up a District Plan, planners may either merely tell people to follow instructions so that they become dependent on instructions every step of the way. Alternatively, planners may work with the local community to try and tackle health problems jointly. This can encourage local initiative and new solutions to old problems (see figure 3.6).

For developing a local strategy of community involvement in health the checklist in table 3.7 may be found helpful.

Figure 3.6   Encouraging self-reliance

Table 3.7 **A checklist for community involvement in health**

1 Have the *objectives* been worked out in open and frank discussions with all sections of the community?
   or
   Were the objectives predefined by the medical hierarchy?

2 Has community participation been encouraged through a process of education and rapport?
   or
   Was partnership bought by coercion or gifts, or a mixture of both?

3 Has there been a genuine transfer of information and skills to the community?
   or
   Is knowledge guarded jealously?

4 Are new approaches and innovations encouraged?
   or
   Are they frowned upon?

5 Is community empowerment the overall aim?
   or
   Is community compliance with all aspects of the health plan the overall aim?

6 Is there give and take in sharing of knowledge?
   or
   Is knowledge being imparted as a form of sermonising or admonishing?

(vii) *Social marketing*

Social marketing comprises a set of strategies related to product, price, availability and promotion to convince the public to adopt a product (for example oral rehydration packets) or behaviour (for example growth monitoring of children, or waste disposal). Community involvement is the cornerstone of social marketing of the District Health Plan for PHC. There are a number of barriers to cross. All PHC interventions are aimed at the poor. They are least likely to be taking a newspaper, or have time to listen to the radio. Hence traditional forms of interpersonal communication must be looked for, for example puppet shows, song-and-dance groups, sermons following prayer meetings, and so on. There is a difference between 'awareness' and 'being knowledgeable' about a subject. In order for people to learn about and understand the important elements of any issue, including PHC, the message needs to be repeated and reinforced in a variety of ways.

(viii) *Matching needs to resources (allocating tasks to resources)*

Needs are often great, and resources limited, so another planning skill is allocating tasks to match available resources. The cycle is shown in figure 3.7 below.

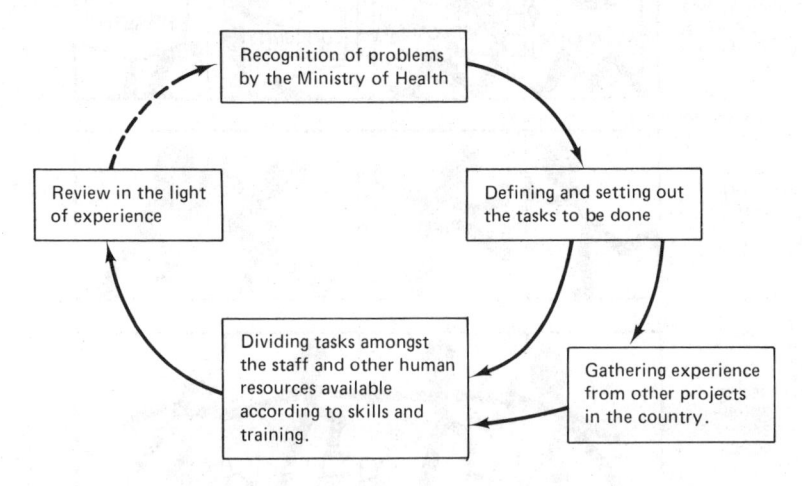

Figure 3 7   Matching needs to resources

(ix) *Liaison and team work*

Effective action on several of the more common health problems requires action in other related sectors besides health. Agricultural productivity is closely related to nutrition. Environmental sanitation is related to prevalence of worm infestations and diarrhoeal disease. Appropriate education is closely related to improved knowledge and understanding of disease causes, disease processes and how effective intervention can be done. Thus liaison between government departments is essential and barriers must be broken down for effective planning (see figure 3.8).

The usefulness of joint planning and collaborative action between sectors is widely recognised, and yet it has been difficult to implement in practice. The problems to be overcome are as follows:

(1) The administrative set-up tends to be rigid and sectoral priorities so very different that pooling and sharing of resources becomes difficult. Vertical relationships with regional and national level superiors continue to dominate.
(2) Imaginative approaches in developing intersectoral strategies for helping the 'high risk' groups are lacking.
(3) The health sector takes for granted that the concept of PHC is widely known, and rarely takes the opportunity for keeping the other sectors informed about the national and District priorities in PHC.

Figure 3.8   Liaison between government departments

Health planning at all levels has tended to remain confined to the health sector alone, carried out principally by health professionals, because of the perception that health of the population is determined by the medical services alone. Such a blinkered view ignores the fact that as a pre-condition for enjoying good health a number of Basic Needs must be met. These are for example:

(1) Consuming adequate amounts of nutritious and safe food to meet physiological requirements. This in turn requires efficient production and procurement of food.
(2) Appropriate shelter and a sanitary environment.
(3) Access to services for maintaining health and psycho-social well being.
(4) Ability to plan the spacing and the number of one's children.

Several of these Basic Needs may be within the capability of the individual families and the community to achieve by themselves. Even then empowering the community to develop local institutions and structures for community action is often needed. Examples of such an approach are Farmers' clubs (to discuss methods of improving agricultural output), Women's groups (to discuss matters related to nutrition and family welfare), and Youth clubs (for leisure activities, health education, and activities for improving the community habitat). Some may require joint action between the District administration and the community. Some require additional inputs from the central government. But unless a local mechanism is created for the community to take a leading role in planning its own development, and for channelling inputs to that effect, all developmental work will be left to chance.

Experience in several countries has shown that the District development committee, and its executive arm, are useful venues for promoting intersectoral action. They provide a forum for informing the other sectors about their roles in the implementation of PHC. Regular workshops and seminars help to merge together the plan objectives of different sectors into a shared vision of PHC. Agreeing local standards for Basic Needs provides not only the framework for intersectoral action but also for identifying those whose health is at risk because their basic needs are not met (see figure 3.9).

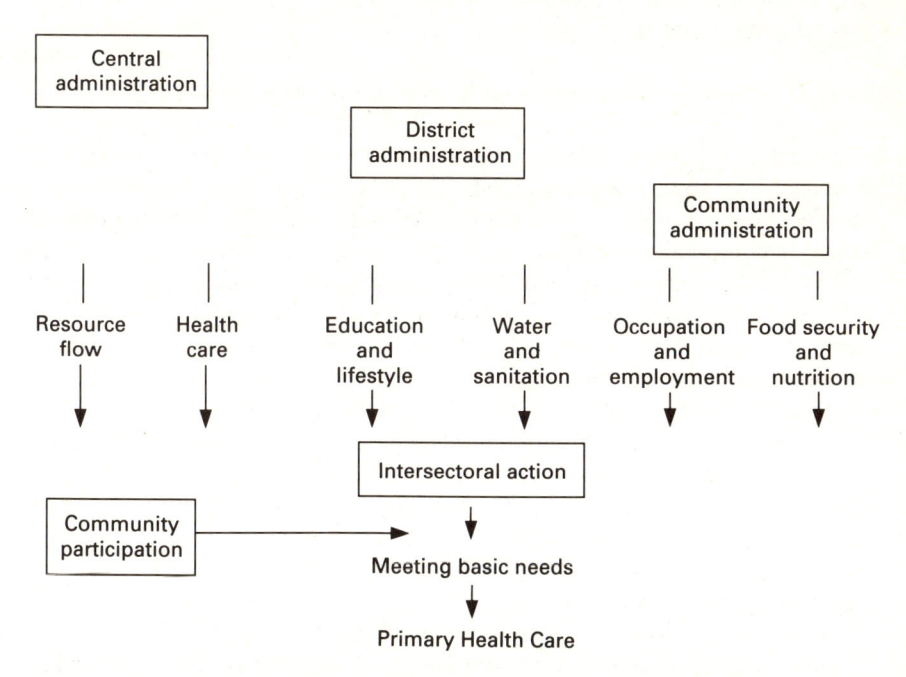

Figure 3.9   Intersectoral action to meet basic needs for Primary Health Care

(x) *Learning from experience of projects, innovative approaches and so on (what goes well, what has gone wrong and what can be done)*

Some of the *relevant experiences* of how things can be done need to be specified using the experience of various projects. This procedure will then need to be complemented by specifying *also where things have gone wrong, and why, in the existing system*; unless these existing problems are overcome, and some solutions found, the same difficulties are likely to recur and be perpetuated into a new system. Much can be learnt from considering what has been successful (or not successful) in the past and why, and what has been found to work elsewhere. A number of new ideas so generated can be utilised for planning. Planners sometimes forget that theirs is never the first new approach to a problem.

(xi) *Setting up a supportive organisation*

Logistic support and operational planning are other key issues recognised as essential in setting up a District Plan. An administrative set-up is needed which will enable individuals to work together, to make decisions regarding actions to be taken, and to supply support and supervision as needed.

Different types of organisation and management are needed in different situations and part of the skill required is to recognise what situations are appropriate for which procedures. Some situations (for example, emergencies) require closed rigid approaches in management. Other situations (most) require an open flexible approach if people's potential is to be fully developed (see chapter 4 and table 3.8).

(xii) *Feedback (data collection and use in local communities)*

Planning is a continuous as well as a learning process involving a series of cyclical steps, and so mechanisms should be built into the process which will ensure that the information required locally, at health units and by the District team, is continually obtained. In the past many health workers have been asked to collect information which they then pass on to others for analysis and without any relevance to decision-making locally. The activity has therefore seemed a waste of time and many health workers have developed the bad habit of thinking that if they merely put in a certain number of hours they will have done their job adequately. *Jobs need to be geared to specific tasks instead of the hours spent at work*. The situation needs to be changed so that health workers can immediately assess for themselves the extent to which they have done a good day's or a good week's work. They have targets to aim for in the short and in the long term. With the involvement of local communities it is obvious that data collection, analysis and interpretation is needed at the very heart of the system, viz. the local community, so that immediate feedback of progress is available and decisions on new actions can be taken locally.

Feedback should also involve constructive criticism from those involved in

Table 3.8 **Types of management**

	Closed (rigid)	Open (flexible)
Data	Enough data to know what should be done and how to do it	Enough to start to learn by doing
Implementation	Using what is known	Avoid using and transmitting erroneous information
Evaluation	How well can we do it?	What is most important?
Changes	Errors and wrong: to be controlled	Expected and useful: opportunities to learn
Planning	Final plans before action	Tentative plan, constant revision during action
Nature of organisations	Unchangeable, internally stable, people come and go	People persist. Organisations constantly change
Authority	In manager with sufficient information to make decisions and get them implemented	In group with enough skills to make good decisions and to remove obstacles to their implementation

Source: Burkhalter, B. R. (c. 1978). *Modeling and simulation* vol. 5. Nutrition Planning Information Service, P.O. Box 8080, Ana Arbor, Michigan 48107, USA.

Table 3.9 **Target setting and feedback mechanisms for five areas of a district plan**

Topic	Targets	Feedback
(a) Financial aspects	Budgeting in advance	Accurate accounting of what is spent
(b) Personnel	Detailed task centred job descriptions	Supervision to improve performance
(c) Supplies	Plan for the quantities required, their procurement and delivery	Monitor use and schedule maintenance
(d) Transport	Budget for mileage and maintenance	Maintain records on mileage and maintenance
(e) Community	Community diagnosis. Identify priority health problems	Community assessment of services, coverage, and utilisation

putting the plan into action. Only by getting such criticism can a plan evolve and develop further. Local communities and local health facilities differ and each will need to work out how best to put the plan into action in their area.

Monitoring of progress also helps in getting feedback (see table 3.9).

(xiii) *Use of key procedures for training in primary health care*

Training is one of the most effective means of implementing a District Health Plan. It should be:

(a) *Focused on problems and tasks*

Training processes usefully include problem orientation, starting with a community diagnosis of priority health problems, developing approaches to the common health problems, trying to anticipate other difficulties and learning from such difficulties as are encountered on the job so that training continuously improves. In this way trainees will become equipped to deal with the type of problems they are likely to meet 'on the job'. Training then becomes *task-oriented*, focusing on the tasks to be done on the job when training is completed.

In the local community, training can be a way of mobilising local resources for health.

(b) *Focused on team work and multi-purpose workers*

Provision of health care relies on many people being able and prepared to work together. This is the common procedure in many societies. It needs to be reinforced. *Training for team work* may be helped by holding joint sessions for various topics. 'People do not learn much from what they are told. They learn from what they think, feel, discuss, see, and do together.'

Teams are needed at the local community level, the Health Station level, and in the District. Team members would have *joint responsibility* for the promotion of health and the prevention and treatment of disease in their area. They would have *overlapping roles* so that one team member can step in for another if someone is away or can cross-refer within the team to more specialised knowledge. All team members are expected to have a *general knowledge of the seven primary health care topics* (child health; maternity care; household environmental protection; community environmental health; community development; curative care and first aid; and communicable disease control). In addition, at each level (viz. the local community, the Health Station and the District), team members would have *specialist knowledge in Household Family Health, Community Environment or Community Organisation for Development* (see figure 3.10).

In starting the training of local community workers it is probably unrealistic to try to train them all to do everything at once. They may not learn anything at all. Experience in different parts of the world indicates that for community health workers an on-going training programme with regular weekly, fortnightly or monthly sessions is more effective than longer courses held at infrequent intervals, or a one-off course.

There is a great need for multi-purpose workers. At the moment huge numbers of uni-purpose workers exist who cannot deal with the multiple

District health plan 111

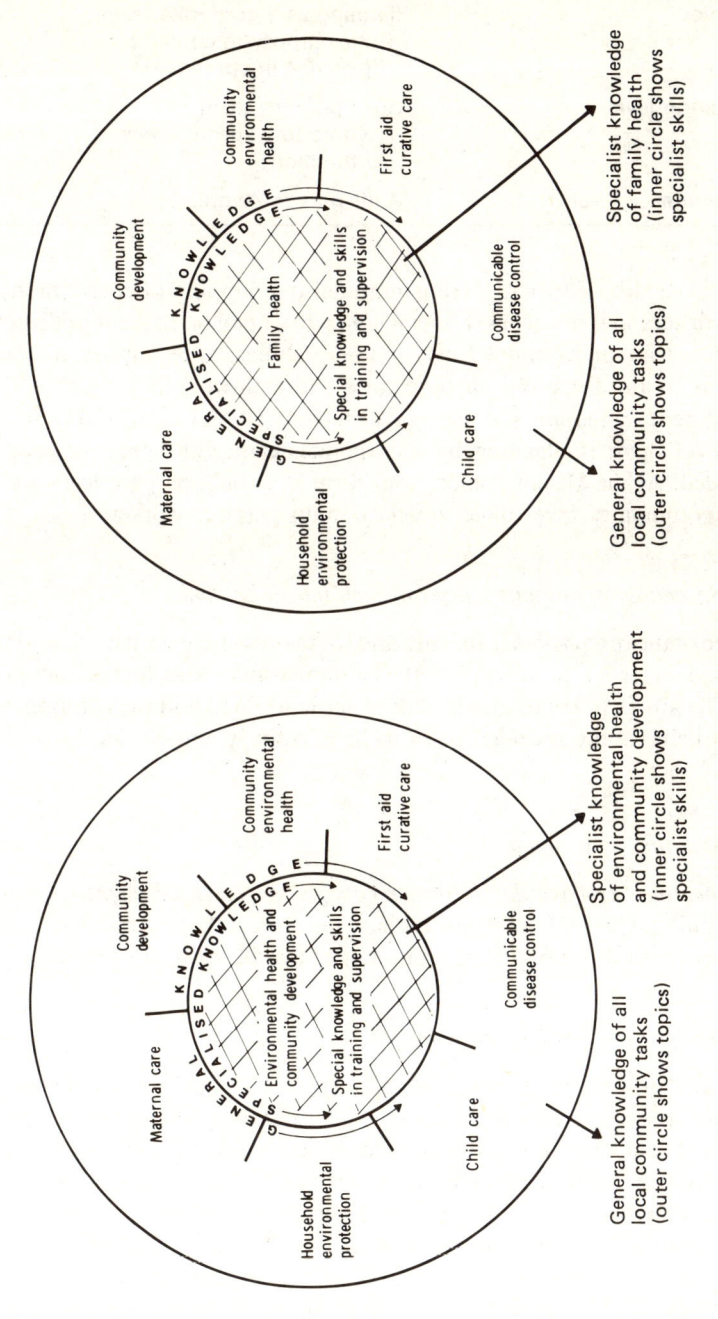

Figure 3.10 Distribution of general and specialist primary health care knowledge

Table 3.10 **Teamwork and multi-purpose approaches**

District level	Planning as a corporate team (Multi-purpose in content; Specialist in approach)
Health Station level	Multi-purpose team (with a few special areas of function)
Local Community level	Multi-purpose workers

causes of ill health. Further, there is minimal overlapping of roles amongst existing workers so if one worker is away, no work at all on that subject can be done. This is an uneconomic way of doing things. One answer to these problems is to aim for multi-purpose roles and teamwork.

The different teams are summarised in table 3.10 and figure 3.11. At the District level there is planning by a corporate team with specialist support when needed. At the Health Station level there is a multi-purpose team and at the local community level there are only multi-purpose workers.

Calculating degree of responsibility for each job to be done

The degree of the *responsibility* for care and for teamwork at each level needs to be analysed in order to work out the administrative and logistic support needed. The aim is to try to identify the exact work load and the components required to enable the specified tasks to be effectively and efficiently carried out.

Regular updating of training

Regular meetings and *regular refresher courses* are essential elements of in-service training. Health knowledge is changing so fast that everyone needs to try to keep up to date. Again, training cycles can be built into the plan.

Interchange between service and training

Another key process to maintain standards of training is the concept of *interchange between service and training responsibilities*. This means particularly exchange between administrative and clinical work at the Health Station, and training work with local community teams. It also means that any full-time District trainers (for initial PHC training) should rotate this responsibility with responsibility for running District services as part of the District Management team. There may also be similar exchanges between some hospital work and some Health Station work. In these ways training and service will remain closely linked together.

District health plan 113

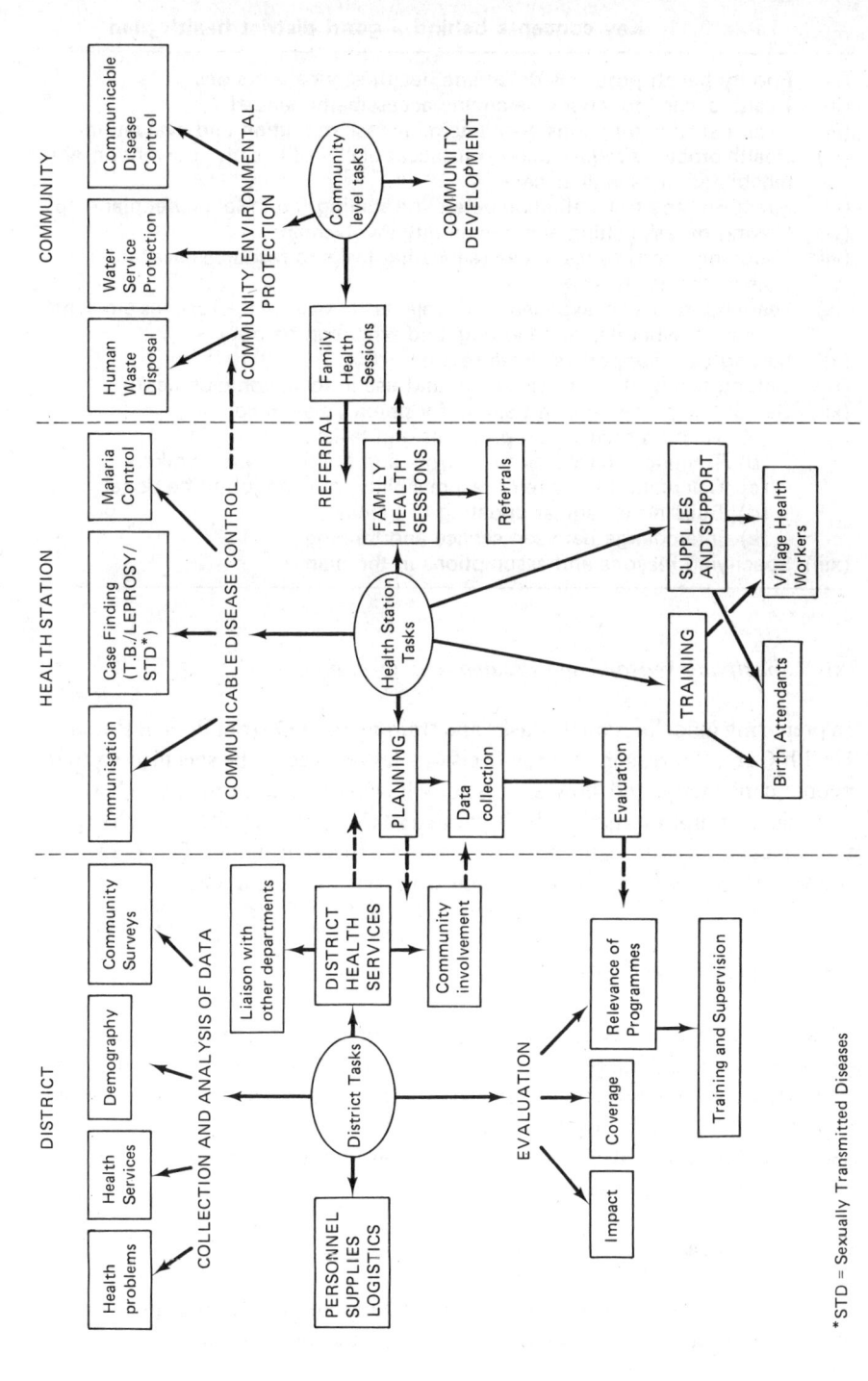

*STD = Sexually Transmitted Diseases

Figure 3.11 Teamwork and multi-purpose approaches

Table 3.11 **Key concepts behind a good district health plan**

(i) Priority health problems determine health service functions
(ii) Health service functions determine accessibility needed
(iii) Health service functions also determine task definition and evaluation
(iv) Health problems require action on causes of ill health, early intervention, and rehabilitation as well as care
(v) Functions and task definition determine staffing (and Manpower planning)
(vi) Linking of task setting and community involvement
(vii) Matching needs to resources (allocating tasks to resources available)
(viii) Liaison and team work
(ix) Learning from local experience of projects, innovative approaches etc. what goes well, what has gone wrong, and what can be done
(x) Setting up a supportive organisation
(xi) Getting feedback (data collection and use in local communities)
(xii) Use of key procedures in training for primary health care
    (a) Training focused on problems and tasks
    (b) Training focused on team work and multi-purpose workers
    (c) Calculating the size of responsibility for each job to be done
    (d) Building in regular updating of training
    (e) Interchange between service and training
(xiii) Specifying reasons and assumptions in the plan

(xiv) *Specifying reasons and assumptions in planning*

In preparing guidelines on the tasks and training for a District Plan of Primary Health Care, the reasons for the decisions taken need to be specified so that people can understand how suggestions have come to be made. Wherever possible, assumptions need to be explained so that when the situation changes, the logical processes can easily be followed through again. The plan is a framework for further discussion and decisions at a local level.

(xv) *Holding reviews of progress*

This is frequently necessary with a complex task, in order to ascertain what proportion of the targets has been achieved. Often, further activity is dependent on achievement of earlier targets, and failure to do so in the apportioned time may delay further development of the health plan. The above key concepts of a District Health Plan are summarised in table 3.11 and a checklist provided in table 3.12.

Resource allocation and budgeting

Plans become a reality when they are translated into budget allocation. Hence resource allocation and budgeting are part of the planning process.

Table 3.12 **Checklist for a plan**

Background	Is the information relevant? Have the needs and priorities been clearly assessed?
Long-term goals	How well do they fit into the overall policy? Are they clear and realistic? Do they give a clear vision of the future?
Objectives	How do the objectives fit into the long-term goals? Is there sufficient commitment for the goals to be achieved? Have all the different alternatives been carefully considered? Can the plan accommodate an alternative strategy if one becomes available?
Action plan	Who has the overall responsibility of the plan? Who has the responsibility for implementation of different sub-units of the plan? What mechanisms exist for co-ordination of different activities of the plan? Does the plan specify key result areas? Are individual staff members specified to take the responsibility for different parts of the plan? Are the results feasible? How much flexibility is possible?
Resources	Have all the resources needed for the plan been carefully identified, e.g. manpower, facilities, equipment and material, community goodwill?
Implementation	How will the implementation commence? What is the order in which different parts of the plan will be put into effect? By whom? Who will monitor?
Evaluation	How will the plan be evaluated? At what point in time? By whom? Have the criteria for evaluation been set?
Balance	Is the plan well balanced with no preponderance of any particular part? How does the achievement of each objective reinforce the long-term goals and the overall policy? Does the failure in achieving any one goal weaken the plan overall?

The budgetary framework and details need to be worked out to reflect the different major undertakings under the plan. When all the expenses and estimates have been carefully worked out it makes the process of negotiating the plan with regional and national authorities that much easier. It is always helpful to have the annual plans and budgets widely commented on and approved by all parties concerned if intersectoral activities, community involvement, and co-ordination with NGOs is to take place. The steps in the planning process will be as follows:

(1) Developing a three-year or five-year District Health Plan.
(2) Writing the District's one-year Forward Plan and priorities.
(3) Translating the Forward Plan into estimates of costs.
(4) Reconciling estimates with financial allocations.
(5) Controlling expenditure in the light of allocations.

In most countries three basic weaknesses can be found with regard to the financing of health plans. These are:

(1) Financial management and accountancy systems are inefficient and cumbersome, with the result that on-going information about how much is being spent on what is inadequate. Hence day-to-day decision making is largely empirical.
(2) There are not enough resources in the health sector, which is also the sector first to be squeezed at times of economic recession.
(3) The available resources are not equitably distributed between curative and preventive care; between health posts, health centres, and hospitals; between recurrent expenditure and new developments.

It is necessary to develop operational budgets which correspond more closely to individual programme objectives instead of having everything lumped together under general headings. Separate accounting systems for each major activity will provide a sound basis for reviewing costs incurred for different activities for example for main PHC programme areas (sometimes referred to as cost centres) like MCH services, communicable disease control, environmental health, curative services, administration, and so on. This is more efficient than having all costs amalgamated and shown under salaries, travel costs, supplies, and so on. Moreover, information on utilisation of services and expenditure can be compared with epidemiological data to assess the performance of each activity. If there are diverse donors supporting different activities the sources of funds for each activity can be shown separately.

Efficiency in the use of resources is necessary for avoiding wastage. If the salary costs of health workers take up 80 per cent of the health budget it is wasteful to have them idle because of lack of drugs, stationery or transport. An essential drugs policy is another way of introducing efficiency in spending in a major item of expenditure. The use of generic drugs, and restriction in prescribing to specific essential drugs depending upon the level of the health facility is a good way of avoiding wastage.

A number of countries are now experimenting with various ways of raising local revenue, either through user charges (for example the Bamako Initiative) or through pre-paid health cards where the possessor of the card is entitled to free preventive care or free treatment for specified illnesses.

Examples of different elements of a district health plan

Four elements of making a District Health Plan are described in examples in this section. These are: (i) Recognising priority problems and relating them to health service functions and tasks; (ii) Planning and implementation whilst recognising what can go wrong; (iii) Planning training procedures and drawing on experience of what can go wrong; (iv) Calculating the degree of responsibility for each job to be done, how many staff are needed and what their requirements are likely to be. The example given is of maternity care.

(i) *Recognising priority problems and relating them to health service functions and tasks at each level*

The first step is to identify those problems which have priority on account of their prevalence, severity, community concern and the likelihood of response to management. The next step will be to decide on services that need to be provided in order to respond effectively to these problems. This is followed by agreeing the tasks which should be carried out for each service. Such a breakdown for maternity care is shown in table 3.13.

The next step is to recognise what is going wrong with existing maternity care services. Every society endeavours to care for mother and child. A District Health Organisation needs to build on what is already being done, and face difficulties where these are occurring. In a study of two Districts which examined all maternity care (amongst other types of care) provided at government and private maternity units, Health Centres and in outpatients at the District Hospital in Ghana a number of important shortcomings in the existing service were found. History taking of previous obstetric experience was very poor (because it was often done by untrained personnel) and height was rarely measured as an indicator of possible disproportion if pregnant women were under 51 in. (145 cm). The 'risk concept' was not used except by a few individuals, so there was little effective screening of high risk cases and referral was not generally based on particular high risk categories. There were shortages of tetanus immunisation, malaria prophylaxis and iron tablets. There was little antenatal health education on home delivery (although many mothers do deliver at home) nor on family planning. In both Districts family planning services were hardly available at all, in spite of considerable demand. When mothers did need to be referred to the hospital in an emergency this was extremely costly. Only one fifth of births were supervised by a trained attendant, but for those that were supervised the standard of delivery care seemed to be relatively good (see table 3.14).

The next step is to identify the defects and deficiencies in the existing maternity care system. Such an exercise is essential because, until existing difficulties are solved, there can be little point in attempting to extend the services further into the community. All new ideas need to begin with the retraining of existing staff.

Table 3.13 **Maternity service functions derived from priority maternal health problems**

Priority maternal health problems	Contributing conditions or risk factors
(1) Causes of maternal mortality	
Ruptured uterus	disproportion*, malpresentation**, previous Caesarian section*, malposition**
Haemorrhage (mostly antepartum haemorrhage (APH), postpartum haemorrhage (PPH))	multiparity*, anaemia**, complicated delivery, abortions, placenta praevia
Puerperal sepsis	unhygienic procedures during delivery
Pre-eclamptic toxaemia	aetiology unknown – indications: high blood pressure, excessive weight gain**, clinical oedema, proteinuria**
(2) Causes of neonatal and perinatal mortality and stillbirths	
Birth injury	prolonged labour, abnormal presentation, disproportion*, etc.
Intrapartum asphyxia	toxaemia**, twins*, past and present malnutrition**,
Low birth weight	malaria**, primigravida*. Age over 40 and under 18*, previous stillbirth*
Tetanus	Unhygienic methods for cutting and tying the cord

* Can be detected at first antenatal visit
** Can be detected or prevented during antenatal period

Source: IDS Health Group, 1978. *Health needs and health services in rural Ghana.* Institute of Development Studies, University of Sussex, UK.

Table 3.14 **Summary of existing maternal care problems in the district**

Data collection problems
History-taking problems
Need for height measurement
Need for tetanus immunisations, malaria prophylaxis and iron tablets
Need for effective screening and referral based on 'at risk' concept
Need for family planning and home delivery health education
Need for low cost emergency referral
Need for more coverage of family planning services
Need for more supervision of delivery by trained personnel

Source: IDS Health Group (1978).

Having recognised the health problems as well as the defects and deficiencies in the existing health care system, and having specified the tasks to be done, the next stage is to allocate the tasks to the available resources (see table 3.15). This is also shown in figure 3.11 commencing with District level tasks, those at the level of the health stations and those to be carried out in the community.

(ii) *Planning implementation and recognising what can go wrong*

One crucial activity is to organise the logistic support and administrative arrangements for the District health service. A checklist of basic logistical support required will include the following:

(1) Existing record systems may need improvement. For example, the antenatal card or the Road to Health Chart may need a 'Risk' section and be action-oriented.
(2) Requirements for transport need to be estimated and budgeted for.
(3) Equipment for the cold chain will also need to be similarly estimated and budgeted for.
(4) An essential drug list should be prepared for the District and circulated.
(5) A drug allocation system needs to be developed based on requirements rather than enlightened guesswork.
(6) A workable accounting and budgeting system needs to be developed.
(7) Laboratory and other services may need regular updating and maintenance.

People work more effectively if they have clear notions of their areas of responsibility and have been trained or given the opportunity to learn about the tasks they have to carry out. It is no use expecting doctors, nurses and other health personnel to come out of their training institutions knowing every detail of the tasks they are expected to perform. A considerable amount of training occurs on the job. There are other similar pitfalls which can affect the success of a District health programme. A checklist of management priorities to avoid pitfalls is given below:

(1) Clarify hospital responsibilities. Curative and surgical work has great attraction. If the staff assigned to hospital work are away, nurses and doctors from the District Health Team must not be moved to hospital duties.
(2) Clarify the work balance of the District Medical Officer (DMO). Up to half the time can be taken up by administration and the DMOs need to be prepared for it. Good and efficient administration is half the battle and if the DMO only does it sporadically and grudgingly, the District health programme will falter.

Table 3.15 **Maternal services: functions and allocation of tasks**

Functions of maternal services	Tasks to be carried out
Ensure normal delivery of a healthy baby and maintain mother in good health before, during and after delivery by:	Ensure (i) A minimum of 1 antenatal contact for all pregnant women (ii) An average of 3 antenatal contacts for 60% of pregnancies (iii) Where there is more than 1 antenatal contact, the first such contact to be in the first half of pregnancy
1 *Antenatal care* for the purpose of: (i) diagnosis of mothers at risk of developing complications (ii) treatment of some complications and referral of others to facilities where more sophisticated management is provided (iii) prevention and treatment of diseases arising or exacerbated during pregnancy (e.g. anaemia, malaria) (iv) nutritional, health and family planning advice (v) protection of the baby from tetanus	*Antenatal care* (i) Take a history to identify the risk factors (ii) Refer those patients with risk factors (iii) Measure height and weight, fundal height and abdominal girth (iv) Check haemoglobin, blood pressure and urine (v) Do clinical examination. Examine abdomen after 7/8 months (vi) Give tetanus immunisation, malaria prophylaxis and iron and folic acid tablets (vii) Give advice on nutrition, checking on foetal movements, and on breast feeding
2 *Supervision of deliveries*: for the purpose of (i) ensuring aseptic techniques (ii) early detection and treatment of complications during delivery (iii) proper care of the newborn	*Delivery care* Supervise all births either directly or through trained village birth attendants (TBAs) (i) Monitor progress in labour using cervical dilatation charts to recognise complications early (ii) Look for signs of complications – foetal distress, bleeding, obstruction etc. (iii) Ensure hygienic techniques during labour (iv) Care of the newborn – ensure normal respiration, aseptic cutting and care of cord. Put the baby on mother's breast (v) Deliver placenta

Table 3.15 (contd.)

Functions of maternal services	Tasks to be carried out
3 *Postnatal care*, for the purpose of (i) early diagnosis and treatment of complications in puerperium (ii) advising mother on newborn child care, nutrition, family planning (iii) provision of contraceptives	*Postnatal care* Check condition of the mother. Look for fever, bleeding, tenderness of abdomen; check size of uterus; ensure adequate breastfeeding procedures. Advise on care of the baby. Give contraceptive advice. *Family planning care* (i) Advise mothers and community on child-spacing (ii) Provide contraceptives

(3) Clarify the DMO's responsibilities for the staff, so that he should have the authority to recruit, re-train, reprimand or recommend as the case may be.
(4) Arrange for regular staff meetings. Job dissatisfaction is likely to arise when health workers become isolated from their colleagues and feel that they are shouldering all the responsibility alone. Several issues may need airing to remove any doubts. Also, the nature of community health activities is such that long-drawn-out and painstaking work and negotiations are often needed, unlike curative work.
(5) Arrange for local collection, analysis and interpretation of data. Such an exercise, carried out on a regular basis, helps the health team to understand the progress being made towards the achievement of the targets set in the health plan.
(6) Plan and arrange for staff training to match the tasks to be carried out for providing health care in rural and peri-urban areas.
(7) Improve staff amenities and provide for more social meeting places.
(8) Rationalise pay differentials. It may not be possible to do anything about salaries but perks can be rationalised. This applies especially to private practice by the physicians and the specialists, which can be restricted to a specified number of sessions per week.
(9) Many of the auxiliary staff may feel that they are in a career cul-de-sac and cannot advance any more. Hence career development of the junior staff requires thoughtful consideration. Encouraging upward mobility creates a climate in which all the staff are working towards furthering their knowledge and being rewarded for it.
(10) Improve the quality of care at the health stations through regular in-service training, preparing a compendium of standardised procedures, ensuring reliable supply of drugs and equipment, and of transport when needed to reach remote communities.

What can go wrong with the health plan?

In trying to set up a new plan it is important to recognise the possible ways in which things can go wrong. At the *Health Station Level* there are particular traps to avoid. The main problem likely to arise at Level B is that inadequate resources will be available to back up the work of the Health Station. In the Primary Health Care system the Health Station staff are the key link between the District team and local community workers. If they are to discharge this responsibility effectively they need considerable support. If the current problems of shortages or inappropriate training at Level B cannot be solved, then it is unlikely that increasing the number of village health workers or trained Traditional Birth Attendants is going to do any good.

Other problems are related to uncertainty concerning the involvement of the community in Primary Health Care. Is it 'outreach' or 'involvement'? There are several grades of participation beginning with nominating leaders to sit on committees to a truly democratic dialogue with the community or its elected leaders. The latter approach is usually more successful in helping communities organise themselves for health as part of over-all rural and peri-urban development.

Procedures for liaison and co-ordination with agriculture, water resource development, education, social welfare, community development, and the District Council, need to be worked out in detail. Special efforts are needed to forge links with extension and field workers who work closely with the local people.

Staff attitudes may cause difficulties at times. There may be unwillingness to work in rural areas; hospital work may be considered more prestigious; the primary aim may be to obtain a higher status in a chosen speciality rather than serve the people; the slant may be towards a high technology approach rather than making use of promotive techniques; and so on. Perhaps one cause of such attitudes is that in the average developing country two-thirds of the trainees come from the urban areas but two-thirds of the postings are to rural areas. Student selection procedures may need careful revision.

A number of things can go wrong at the *community level* in the working of a District Health Plan. In particular the following need careful consideration (see figures 3.12 and 3.13):

(1) *Confusion between contribution and true involvement.* In the environmental or development activities people may be asked to contribute in cash or kind (material or labour). This is different from involvement, and is not a measure of participation. Involvement is to do with deciding about goals and tasks and working together to achieve this.
(2) *Involvement of local people.* Although in theory participation of the people may be intended, in practice it may not take place. Community organisation and the formation of various social groups may be needed to achieve a high degree of awareness and sharing of responsibility.

District health plan 123

*SWCD = Social Welfare and Community Development

Figure 3.12 A 'top down' health care delivery system with a noticeable lack of co-ordination with other sectors, and little or no community involvement

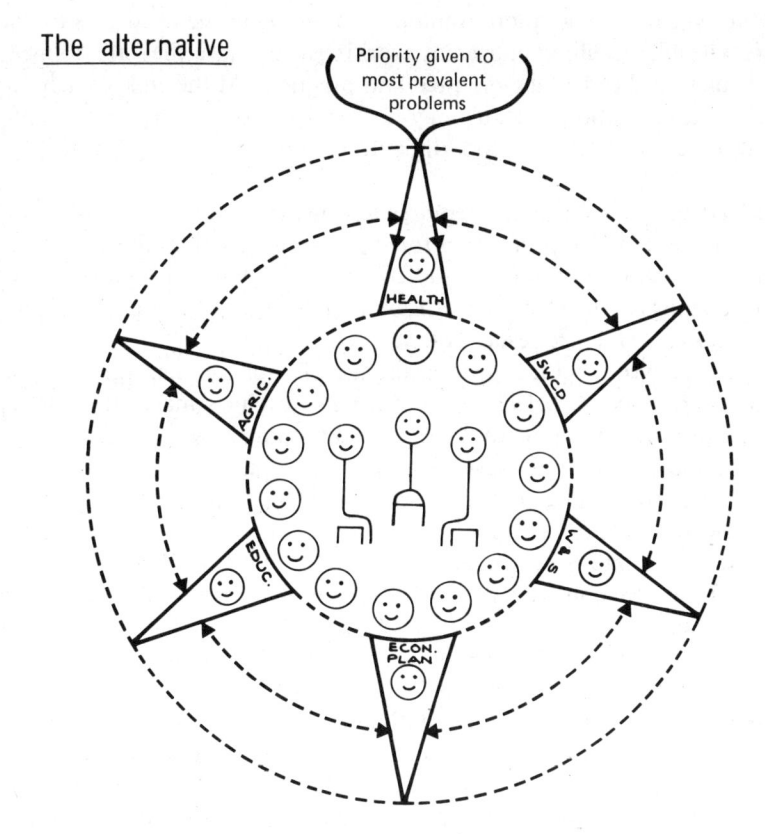

The alternative

Figure 3.13 Community involvement in planning and implementation

(3) *Over-emphasis on structures.* Activities for generating community cohesion may get diverted into a programme of public works, or of amenities and ultimately end up as an administrator's programme.
(4) *Communities where contributions are too few or communal labour difficult to organise* may be neglected. These are usually the poorer communities where the struggle for survival is hard and leaves very little margin. Disappointments and frustrations may lead to the neglect of such communities.
(5) *Excessive decision-making by officials.* Instead of a framework of community health activities evolving out of community diagnosis, health care may be dispensed from above.
(6) *Continuing support may not be available.* It is necessary to maintain the enthusiasm generated at the commencement of the programme through regular feedback and involvement.

(iii) *Planning training procedures*

The main issue in setting up programmes of training in the District is to find teachers who like teaching and enjoy working with those they are going to teach. Topics need to be task-oriented and practical. At the end of each day synthesis and evaluation sessions may be arranged so that the trainers may assess whether a subject is too difficult or too threatening to the job expectations of the trainees.

At the District Level training requires not just the setting up of the initial training courses but also procedures for supervision as well as the continuing education of Health Station Level staff. This can be done through weekly or monthly meetings at the District centre, annual three-day visits to each Health Station, and an annual refresher course.

What can go wrong with District training? The staff-student ratio may be too low, there may be a lack of appropriate teaching material, trainees' experience may be ignored on the course, field training may contradict formal classes, and incentives for future work may be inadequate. There may be drop-outs of trainees and certain people may contradict the goals of the programme. There could be inadequate support of Level B workers by local community teams and inadequate governmental resources to maintain effective supervision. Insufficient time may be allowed for supervisory procedures and some staff, such as elderly dressers, may find it very difficult to learn. In all situations the need is to identify the problem precisely, to find out the reasons it has occurred, and to try and take action on the problem (see table 3.16).

Introductory sessions for training in the short and long term could usefully include case study descriptions of current local community work in the country. Training of the family health workers needs to be task-oriented so that they learn in turn how to plan a teaching and supervision programme for local community workers, how to keep records of local community supplies,

Table 3.16 Summary of guidelines for training family health workers at the level of the health station (Level B)

1. Essential to have discussions with existing health unit personnel in devising course curricula.
2. Criteria for selection of staff could reflect the jobs expected from them, possibly using fictitious case studies to assess ability to solve common day-to-day problems.
3. Useful to have discussions with potential trainees prior to commencement of the training.
4. Use experience from other projects on selection and training for family health.
5. List 'tasks' to be covered in training.
6. Select teaching methods emphasising case studies and role play so that trainers learn by example those methods which are best for teaching local community workers.
7. Select relevant teaching material including texts and audio-visual aids.
8. Use the experiences of others regarding what can go wrong in training. Which solutions suggested are feasible for implementation locally?

how to liaise with others involved with local community workers (for example, other members of the Health Station team, referral midwives, agricultural field assistants, and so on) and how to evaluate local community training programmes. The teaching methods will need to make use of case studies and role playing so that during their studies the Health Station staff learn by personal experience the best methods for use in teaching Traditional Birth Attendants and other local community health workers (see tables 3.17 and 3.18).

Table 3.17 Training the trainers of community health workers (Level A) needs to be task-oriented to include the following

1. Describing the tasks for community health workers and discussing plans with local chiefs, the village development committee and other representatives of the community.
2. Identifying ways local communities could support community health workers (CHW).
3. Describing criteria for the selection of CHWs so that local people could find suitable nominees.
4. Seeing CHW nominees and discussing their job with them.
5. Making an over-all plan for CHW training.
6. Planning the training of CHWs, if necessary in several rounds.
7. Planning on-going monthly supervision to cover one CHW task per month.
8. Teaching CHWs by demonstration and role play.
9. Keeping records of CHW training, supervision, supplies and problems.
10. Organising supplies for CHW work.
11. Liaising with other Health Station workers.
12. Evaluating a CHW programme.

Table 3.18 **Teaching methods need to emphasise case studies and role playing so that the trainers of CHWs learn by experience the methods that are best for teaching CHWs**

Examples
(a) Teaching methods might include accompanying trainees around the local community and asking them to spot 'nuisances', explaining what can be done by the community within its resources to prevent them.
(b) Another training procedure might be to arrange for trainees to meet relevant key personnel e.g. community development workers, agriculture extension officers, school teachers, etc.
(c) It would be essential to demonstrate simple construction skills for environmental improvement, perhaps at each of the Level A local communities.
(d) Organisation of communal labour might be taught through role playing to demonstrate what can go wrong.
(e) Each trainee might make a map of his own area, marking current environmental health problems and attaching a list of needs for community development.
(f) Demonstrations might be arranged at the trainee's own home and farm to show how food production can be improved.
(g) One session might be spent preparing a supervisory checklist from the CHW tasks.

What can go wrong with community health worker (CHW) programmes?

Studies in several countries have identified the following potential weaknesses in CHW programmes:

(1) *Programmes are often vertically organised.* At the time of planning only a few top level policy makers were involved. Dialogue and consultation with those who became trainers and supervisors at the local level may have been minimal or absent. The result is that much stress has gone on the content of training, and little thought given to follow-up supervision or integration into the existing health system.
(2) *Professional interest is minimal,* because of lack of their involvement in planning. Hence CHWs have been largely used as aides in clinics and health centres rather than as change agents in the community.
(3) *Social and political factors have not been taken sufficiently into account in planning.* CHWs are members of communities which are highly stratified along class, caste, clan or kinship, and other divisions. These are potential sources of conflict. Such divisions affect CHWs' loyalties and positions, and demands made on them. Unlike the employees of the formal health sector CHWs are not defended by a governmental system. The local political climate in which the CHWs operate greatly affects what they are able to do.
(4) *Lack of support and supervision.* Most supervisors are from Level B

and are usually overworked, or not interested in supervision, or not trained in it. Often transport may not be available.

(5) *Functions of the CHWs may be poorly defined.* Sometimes they are given only health promotion tasks, whereas the communities may be looking for curative care nearer home. Sometimes they are trained exclusively for curative care with minimum of health promotion skills.

There can also be other well known weaknesses in the programme like poor selection, lack of motivation, poorly trained teachers, irrelevant curricula, lack of suitable teaching materials, and so on.

It is useful to bear in mind that CHWs have service functions (curative, promotive, rehabilitative, and so on) as well as development functions. The District health system is the framework for the support which CHWs must have to function effectively.

What can go wrong with the training of family health workers at health stations?

(1) The trainers may not know how to teach community and family health workers. They may not be interested in teaching them.
(2) Supervisors may not know what to do.
    A suggested list of tasks for supervisors might include the following:
    (a) Re-establish friendly rapport with community health workers.
    (b) Discuss any problems they may have experienced in their practice or with referrals.
    (c) Discuss any complications or deaths that have occurred since the last visit.
    (d) Collect and record data on activities since the previous visit.
    (e) Periodically evaluate the quality of work by observing the community health workers on their routine rounds.
    (f) Replenish supplies (if this service is provided).
    (g) Educate and update the knowledge of community health workers on one topic per month where necessary.
    (h) Inform the community health worker of further activities such as refresher courses.

(3) Trainers may have difficulty preparing teaching material and record forms.
(4) Trainers may fail to form good relationships with nearby hospitals and Health Centres so that they are unprepared when asked to help with training or when they suddenly begin to receive referrals. Involvement of all participating institutions and sharing of information can help to avoid such difficulties.

Why do things go wrong with training?

The reasons for problems in the training programmes for health workers are many. Several of these problems stem from a lack of commitment to teaching found in many training institutions. In medical school, for example, teaching is often a secondary or tertiary activity after research and clinical work has been done. Secondly, lack of interest in teaching may arise from the very little time allocated to discussing teaching methods and organisation of training programmes in the medical and nursing curricula. Thirdly, promotion may depend on every other activity except teaching.

Another problem in training may be the separation of training from service delivery so that a great deal of the teaching tends to be theoretical. Without any first-hand field experience the students' first encounter with real problems is only after qualification.

Members of the health team who will have to work together are at present trained in separate institutions. There is a built-in separation of roles rather than attempts to hold people together through combined sessions. All these problems are often compounded by the large social gap between teachers and students, as well as between the different categories of health workers. This social barrier may exclude the possibility of interchange of ideas, problems and solutions.

The image portrayed by teachers will be very quickly adopted and emulated by the trainees. This is the so-called hidden curriculum. If teachers are only concerned with their status, think upon hospital curative work as the only health care that counts, are not innovative in their approach, and think that patients and students are annoying interferences in their daily life, then the trainees will also develop these attitudes.

(iv) Calculating the degree of responsibility and the workload for each job to be done and how many staff are needed

With some basic information about the community (for example, 20 per cent are children, the birth rate is 50 per 1000 population) it is possible to calculate the degree of responsibility held by each worker serving a particular population. An example is shown of the expected needs for child care in a community of 500 people, based on data from several studies (see table 3.19).

With such information, the degree of responsibility can also be calculated for the Health Station and District levels, for maternity care, adult sickness and environmental health as well as child care. The expected District responsibility for environmental health and community development is shown in table 3.20 in more detail, with a summary of responsibility for District level family health workers in table 3.21.

The need for a District Health Plan formulated through discussions and by consensus within the District Health Team is obvious for giving direction to

District health plan

the health activities within the District. Without such a direction towards defined objectives, health facilities will continue to perform routine tasks without making any tangible progress. But planning by itself is not enough. A true assessment of the health problems in the District is the first step towards successful planning. Attainable targets and objectives need to be defined and the health resources of the District need to be matched with the problems and the defined targets. The health personnel constitute the crucial part of the

Table 3.19 **Expected needs for child care in a community of 500 people**

No. of children under 5 (20%)	= 100
No. of households	= 63
Weighing of 100 children monthly	= 1200 weighings per year
Family health sessions monthly	= 12 per year
Child sickness (5 contacts per child per year)	= 500 sick child visits per year
Child referral to Level B (20%)	= 100 referrals to level B per yr
Child referral to District care (10% from Level B)	= 10 referrals to level C per yr
Child adms. to District Hospital (25% of referrals to Level C)	= 2–3 children per yr hospitalised
Meetings with other local comm. workers (monthly)	= 12 per year
Meetings with Level B staff visiting Level A (weekly)	= 52 per year
Meetings at Health Stations (monthly)	= 12 per year
Meetings (bi-monthly) at Level A for home visits etc.	= 6 per year

Table 3.20 **Tasks in the district for environment health and community development**

(i) *Tasks*

For 20 health station teams each with 10 local community teams, total 200 local communities, maximum 100 000 population, if *all* tasks were to be done, the district community health responsibility needed would include:

200–400 water sources	(1–2 per 500 people)
200 public latrines	(1 drophole per 40 people, 2500 dropholes)
200–400 refuse dumps	(1–2 per 500 people)
200 market places	(1 per 500 people)
200–400 chop bars	(1–2 per 500 people)
200–400 school food traders	(1–2 per 500 people)
200–400 drinking bars	(1–2 per 500 people)

Table 3.20 (contd.)

? 200 community farms and gardens	(1 per 500 people)
200 community day care facilities	(1 per 500 people)
200 local income earning activities	(1 per 500 people)
1200 community labour activities	(6 per local community per year, minimum)
200 community projects needing district advice once or twice	(2 per local community per year, minimum)

To be estimated: CDC incidence, spraying, mass campaigns and individual care needed.

TOTAL 3400–4400 ongoing activities in all

However, if most Health Workers were doing only the top priority tasks, only some of these projects might be in operation.

(ii) *Area: 25 mile (40.2 km) radius, 1963 sq. miles (5076 sq. km)*
The area covered by the District team would be not more than a radius of 25 miles (40.2 km) from the District base (to enable a round trip of 50 miles (80.4 km) maximum). The maximum total area would be 1963 sq. miles (5076 sq. km).

(iii) *Population: 100 000 people maximum per District team*
It is assumed that one District team could be responsible for a maximum population of 100 000 people (see workload below).

With the distribution of responsibility for 100 000 population into 20 health station teams (1 team per 5000 people) and 200 local community teams (1 team per 500 people) the expected community health responsibility and workload would be as follows (if the distribution of the responsibility for the 100 000 people was different some aspects of the workload would of course differ too).

(iv) *Expected local community Level A workers per District team (100 000 people)*
(In low density areas one Level A worker might be providing top priority tasks only for all three areas, community health, maternal care and child care).

200 community health organisers (1 CHW for top priority tasks for 500 people)
200 retrained TBAS ⎫
200 household family ⎬ which the district MCH supervisor will need to supervise from the health station.
health workers ⎭

Expected health station (Level B) MCH workers per District team for 100 000 people)
20 health station MCH workers.

CDC = Communicable Diseases Control
CHW = Community Health Worker
MCH = Mother and Child Health
TBA = Traditional Birth Attendant (also Trained BA)

Table 3.21 **Summary of responsibility for district family health specialist serving a population of 100 000 people**

No. of family health workers	= 200 (1 per 500 people)
No. of traditional birth attendants	= 200 (1 per 500 people)
No. of community health workers (for monthly supervision)	= 200 (1 per 500 people)

Maternal care

No. of households	= 12 500
Expected number of births	= 5 000 per year
Deliveries by TBAs (?60% births)	= 3 000 per year
Deliveries by Health Station staff (30% births)	= 1 500 per year
Standard antenatal contacts for TBAs (3 per pregnancy)	= 15 000 per year
Standard antenatal contacts for Level B staff (2 per pregnancy)	= 10 000 per year
At risk antenatal contacts for Level B staff (20% pregnancies) 3 extra contacts	= 9 000 per year
Emergency in labour referral to Level C (10% pregnancies)	= 500 per year
Postnatal care by TBAs (3 per birth)	= 15 000 per year
Postnatal care at the Health Station	= to be estimated

Child care in the District (per 100 000 population)

Number of children under 5 (20%)	= 20 000
Number of households registered	= 12 500
Weighing of 20 000 children monthly	= 240 000 weighings per year
Family health sessions monthly (200 communities)	= 2 400 per year
Child sickness at Level A (5 contacts per child per year)	= 100 000 sick child contacts
Child sickness referred to Level B (20%)	= 20 000 sick child events referred per year
Child sickness referral to Level C (District) (10% of those at Level B)	= 2 000 per year
Child admission to District Hospital (25% of those referred to the District level)	= 500 per year

Household health care

Average household size = 8 people
Number of households = 12 500

12 500 kitchens and kitchen hazards
12 500 water storage systems
12 500 household faeces disposal methods
12 500 waste water seepaways
12 500 food storage methods and vermin hazards
12 500 sleeping places potentially hazardous for spreading tuberculosis and other communicable diseases
(??) incidence of communicable diseases

Table 3.21 (contd.)

Adult sickness care per 100 000 population in district

Number of adults over 15 (55% population)	= 55 000
Number of households	= 12 500
Family health sessions monthly	= 240 per year
Adult sickness (2 contacts per adult per year)	= 110 000 per year
Adult referral to Level B staff (20%) (to family health session if possible)	= 22 000 per year
Adult referral to Level C (District care) (10% of those referred to Level B)	= 2 200 per year
Adult admission to District Hospital (25% of those referred to Level C)	= 550 per year

health resources of the District. Their training, deployment and technical as well as administrative support at all levels is a major challenge in health management. In the absence of such back-up and support of the workers even the most carefully designed plans cannot succeed. Hence the importance of building up an effective health organisation which will ensure that the right personnel are deployed at the right level and regularly receive the material and equipment to deal with the tasks set for that level. The key concepts and principles of evolving an effective District health organisation are considered in the next chapter.

WRITING PROJECT PROPOSALS

Preparing the District Health Plan, however important, has to be within the policy guidelines set out by the Ministry of Health. It invariably deals with current health policies and is restricted to existing health activities. These are tightly controlled, being dependent on the release of funds by the Treasury. There is usually very little scope for breaking new ground. On the other hand, many donor agencies, charities, and aid organisations exist for supporting innovative projects and initiatives. There are several examples of new approaches in health care first being demonstrated as workable in small-scale projects before being adopted nationally. District Health Managers need to develop the skills in writing project proposals, and indeed, have several at hand concerning the more pressing issues for presenting to prospective donors at an opportune moment. The pages that follow take the reader through the steps of project formulation and writing a detailed proposal.

Writing a project proposal requires thought and discussion. It can never be a hurried exercise, and normally progresses through three stages viz.

(1) project formulation;
(2) getting feedback; and
(3) writing a detailed project proposal.

Project formulation

Where do ideas about projects come from? The health manager might have recognised an unmet need, or a suggestion came up during discussions and meetings with colleagues, or community groups. A solution for a current problem might have been thought of by a team member. Whatever the source of the idea, the process begins with putting down on paper the propositions and their analyses. At this early stage it is necessary to pause for reflection and critical analysis.

The next step is to write a brief (not more than five pages) Preliminary Project Paper for internal discussion and for canvassing feedback. This provides an opportunity for setting out in a logical manner the results of one's critical thinking, and for describing the need for a project, the goals to be achieved and the strategy to be used. Developing a project paper in writing allows one to be more critical, and to identify gaps and inconsistencies in one's thinking. The most important issue is whether the proposed action or intervention is feasible within the limitations of available manpower and resources. At this stage a rough calculation of available resources and those to be requested from the donor is essential.

Getting feedback

The Preliminary Project Paper is circulated to colleagues and superiors for comments. It also helps to get feedback from the prospective donors as to whether the project falls within their remit and stands a chance of being considered favourably.

Feedback received from one's colleagues and superiors is useful in judging not only the technical quality of the project but also serves as a measure of the organisational and political support the project is likely to receive. In the real world of internal politics and hidden agenda gaining the support of one's colleagues and superiors sometimes turns out to be more arduous than that of the donors. However, in seminars and discussion groups and by lobbying key persons the necessary support for the project can be built up.

In dealing with donors it is useful to remain well informed about their priorities, the size of the projects they normally fund, the proposed formats and procedures they require and the frequency of evaluation they demand.

Even though the Preliminary Project Paper might be used to sound out the prospective donor, the final Project Proposal invariably needs a covering letter from the administrative head. This may turn out to be the District or Regional Commissioner, or a central ministry, or several ministries. Hence the need for keeping one's superiors informed and following their counsel.

Donors have four major concerns, and it is often wise to address these in the Preliminary Project Paper as well as in all negotiations with them. These concerns are as follows:

(1) Is the project economically feasible? (In other words has the applicant done all his sums correctly?)
(2) Is there institutional capability with regard to technical know how, logistics, accounting and managerial inputs? (In other words is the applicant biting off more than he can chew?)
(3) How will the outcome be measured? (In other words has the applicant been doing only wishful thinking and not considered realistic measures of outcome?)
(4) Is cost-sharing envisaged? (In other words are other donor agencies being approached? If so how will they be co-ordinated?)

Writing the detailed project proposal

Having canvassed opinion about the project and carefully laid the ground for it, the next step is to write the detailed proposal. In doing so one should first of all take into account all the feedback information received taking care that the balance between objectives, strategies, inputs and results carefully worked out in the Preliminary Project Paper is not upset. Two topics require careful attention viz.

(1) In writing the Preliminary Project Paper the background information may have been scanty. Now is the time to gather detailed information, all fully validated, in the Project Proposal.
(2) Feedback from colleagues and discussions with them, with superiors and with donors helps to define the organisational and political realities. Full account must be taken of these in writing the comprehensive document.

Anatomy of the proposal paper

Many donor agencies have their own application forms to ensure uniformity. By and large they have a similar format and ask for the same information even though the terminology used may differ slightly. The common layout is as follows:

District health plan

Summary

What is it all about?
This is the most critical section. Decision-makers are busy people who have to read and understand many documents. A well argued, informative but concise and interesting summary is half the battle. The summary must convince the reader that the project is relevant to the aims of the donor agency, and to the needs of the country. It should read as well thought out, and must cover all the essential points.

The summary should include the overall objective of the project, the strategy to be employed, how the project will solve an urgent problem or an unmet need, and the total cost.

Since the summary is an important part of the document it is best to write it at the end (even though it comes at the beginning of the document).

Problem statement

Where are we?
This section covers the background and provides an analysis of what problem is to be addressed.

A *background* section should describe those aspects of the general situation which are relevant to the problem, and to the proposed intervention.

Then follows a section *analysis*. It sets out the importance of the problem, its magnitude, its relevance to the country's health priorities, and its importance for delivery of health care. Whilst writing this section the following points should be borne in mind:
- How did the problem come to be identified resulting in a project to address it?
- Is the problem growing in magnitude?
- What has been done so far to deal with it?
- Any success? Any failure?
- Was any evaluation carried out of previous interventions?
- What evidence is there of a workable solution?
- Is addressing the problem relevant to the nation's stated priorities?

Goals and objectives

Where do we wish to go?
This section should outline the logical link between the problem stated in the previous section and the strategies to be employed described in the following section.

Goals are general statements about what is to be achieved. *Objectives* are more specific measurable targets. Because most problems one tackles are complex, there is usually a hierarchy of objectives. But all should clearly specify the status to be arrived at on the completion of the project. (Donors are particular about the End of Project Status – EOPS.)

This section should address the following issues:
- The overall goal of the activity of which the project is a part.
- What will be achieved by the project?
- What will be the status of the activity at the end as a result of the project.

Strategy *What route will we take?*
This section is at the heart of the project design. The strategy may consist of one or more interventions, occurring simultaneously or sequentially. It is desirable to describe the strategy as a whole, as well as for the individual components. The *assumptions* to be made about conditions to be satisfied before interventions can be carried out, the *methods* to be employed, and the resulting *outputs* need to be set out. The donors are specifically concerned to know how much of the activity will continue after the duration of the project. This may require a description of the existing or new infrastructure, the ability to meet recurrent costs, and any possible obstacles to the successful completion and continuation of the activity.

Implementation plan *How will we travel?*
This is another important section. It describes the details of the interventions to be undertaken. The *inputs* in terms of manpower, equipment, supplies and operating costs need to be specified. Existing resources already in place, whether provided by the government or other donors, should be described, including a statement about how they will fit in with the proposed project.

A *workplan* outlining activities, targets and schedules should be given. A large project with several components may need a master workplan as well as component-specific plans. The purpose of the workplan is to allow easy review of achievements and constraints. Table 3.22 is an example of a workplan. A workplan should indicate the following:

Timing – whether an activity is continuous, intermittent or at specific times.

Table 3.22 **Example of a workplan for baby friendly hospital**

Activity by component	Timing	Objective or target	Critical assumptions	Indicators of achievement
Component A (Writing policy document)				
Activity A1				
Activity A2				
Activity A3				
Component B (Training health staff)				
Activity B1				
Activity B2				
Activity B3				
Component C (Health education)				
Activity C1				
Activity C2				
Activity C3				
Component D (Establishing routines in maternity ward)				
Activity D1				
Activity D2				
Activity D3				

Target – the direct result or outputs of an activity, and how they relate to the objectives.
Critical assumptions – the essential conditions which need to be present before an activity can take place.
Indicators of achievement – the evidence to verify that the activity has been completed and the extent to which the target has been achieved.

It is essential that the workplan clearly indicates the organisational framework within which implementation will take place. An organisational chart showing linkages between the project and other units of the organisation as well as external bodies is always helpful. It tells the donors as to what staff will be assigned or recruited, who will be in charge of the project, and what are the lines of responsibility. It is also necessary to specify whether there will be a steering or advisory committee, its membership, and the terms of reference.

Monitoring and evaluation

How will we know when we arrive?
It is advisable to work out a system of monitoring and evaluation at the outset when the project is being designed. Two kinds of indicators are needed viz. those for reviewing progress against planned activities or targets (that is for *monitoring*), and those for overall review of the results and the process including project design and its relevance to objectives; relevance of the strategies employed; and relevance between implementation and outcome (that is for *evaluation*).

Monitoring is carried out on a day-to-day basis by the project manager and supervisors, by means of field visits, through regular meetings of the project team, and by regular reports. Hence a good workplan is a great help. Evaluation is commonly done mid-term, and at the end by outside consultants.

Costs

What will be the cost?
Costs are usually divided into two broad categories: capital and recurrent costs. Enumerating capital costs is usually straightforward. For recurrent costs, *all* activities including seminars, workshops and evaluation exercises should be costed giving in detail the assumptions underlying the calculations. Allowance needs to be made for inflation and for unforeseen expenditures (contingency funds). The contribution made by one's department to the overall costs also needs to be shown. A budget checklist will comprise the following:

Capital expenditure:
Buildings (offices; housing; clinics; lecture rooms etc.).
Fixtures (furniture; equipment; vehicles etc.).
Recurrent expenditure:
Personnel (technical staff; support staff; consultants; casual labour).
Training (fellowships; study travel; workshops and courses; correspondence; stationery).

	Travel costs (fares; per diem for students and lecturers). Supplies (medical supplies; training supplies; office supplies; printing and photocopying; vehicles). Maintenance (equipment; furniture; building). Vehicle running and maintenance. Other costs (water; electricity; postage; telephone; rent etc.). Contingency. Inflation.
Description of the department	When an application is being made within the country it is likely that the department is known to the donor agency. Even then, and always when the application is to be judged abroad, a description of the department, its staff, the work being done, and the track records of the employees is needed.

FURTHER READING

Gish O. *Guidelines for Health Planners. The Planning and Management of Health Services in Developing Countries. TRI-MED*, London, 1977.

Oakley P. *Community Involvement in Health Development. An Examination of Critical Issues.* WHO, Geneva, 1989.

Rifkin S. B. *Community participation in Maternal and Child Health/Family Planning Programmes.* WHO, Geneva, 1990.

UNICEF/WHO Joint Committee on Health Policy. *National Decision Making for Primary Health Care.* WHO, Geneva, 1981.

World Health Organization. *Formulating Strategies for Health for All by the Year 2000.* WHO, Geneva, 1979.

4 Building the Health Organisation in the District

The District Health System is the practical framework through which the District Health Plan will be implemented. The prime responsibility of the District Health Manager is to set the course and steer the health plan to its agreed targets. Inevitably, the course turns out to be hardly linear. As problems and difficulties arise the course becomes increasingly zig-zag. It is the skills and personality of the District Health Manager which would ensure progress in the general direction of the plan targets.

The District Health System is a self-contained segment of the National Health System. It comprises a well defined population living within a clearly demarcated administrative and geographical area, and includes all institutions and individuals providing health care in the District. Such institutions are principally from the formal governmental health sector, but there may also be non-governmental, private sector, or traditional sector practitioners providing health services of various types (see figures 2.3 and 2.4). Thus the health system comprises several elements which contribute to health in homes, schools, workplaces and communities. The components of the system include all health facilities and the staff working in them (Levels A, B and C) up to and including first referral level hospital. In the implementation of the health plan the challenge for the District Health Manager is to draw together all the component elements of the system and co-ordinate their activities in the direction of the promotive/preventive, curative and rehabilitative targets set.

District Health Systems have vertical relationships with higher level regional management, horizontal relationships with local departments of other ministries, between different national health programmes, and with the communities that are being served. Depending on the political system there may also be a variety of horizontal relationships with local political and community organisations. Thus there is plenty of routine administrative and service work to keep everyone occupied. Without a guiding hand the systems will remain engrossed in routine work with little progress forward. Moving forward in the direction of the plan targets whilst continuing to maintain the standards of daily routine work requires skilful management. This explains why some Districts do better than others.

As mentioned earlier the District Health System rests on five pillars. Reference has been made to them in other chapters, for example:

(1) Organisation, planning and management (page 86)
(2) Financing and resource allocation (page 114 and 242–248)
(3) Intersectoral action (page 105)
(4) Community involvement (page 102)
(5) Development of human resources (page 100)

Probably the most important instrument available to a District Medical Officer (DMO) in charge of the District Health Team for implementing a Primary Health Care Plan is the organisation of the team. One of the marks of the DMO as a successful manager is the ability to visualise the health personnel and other resources of the District as an organisation, disposed in the most effective way throughout the District to combat its health problems. Just as the health plan gives a clear analysis of the health needs of the District and the most effective way of meeting them, so the organisation is the means of translating the objectives of the health plan into practical action. Where there is good organisation the health personnel are able to work to their full potential, and the many different aspects of health care carried out by different people can be integrated for maximum effect. It is only within a sound organisation that good management can take place. There are many examples of individuals whose work has been unsatisfactory but which has improved enormously when they have either been moved to a different organisation or to another part of the same organisation, or the organisation itself has been changed in some way. What has been at fault is not the individual, but the organisation.

THE HEALTH ORGANISATION AS AN INTERNAL SYSTEM WITHIN AN EXTERNAL ENVIRONMENT (see figure 4.1)

The health infrastructure of the District comprises the health organisation of the District. An organisation has its own internal systems which enable it to deal with the demands and pressures of the external environment. In the case of health the external environment consists of health needs and priorities, the felt needs of the District population, national policies, political pressures, economic constraints, local politics, and so on. The internal systems comprise those related to knowledge, skills and technology as well as transport, supplies and communications. The internal systems of the health infrastructure enable the health organisation of the District to deal effectively at its interface with the external environment. The more robust and efficient

142   *District Health Care*

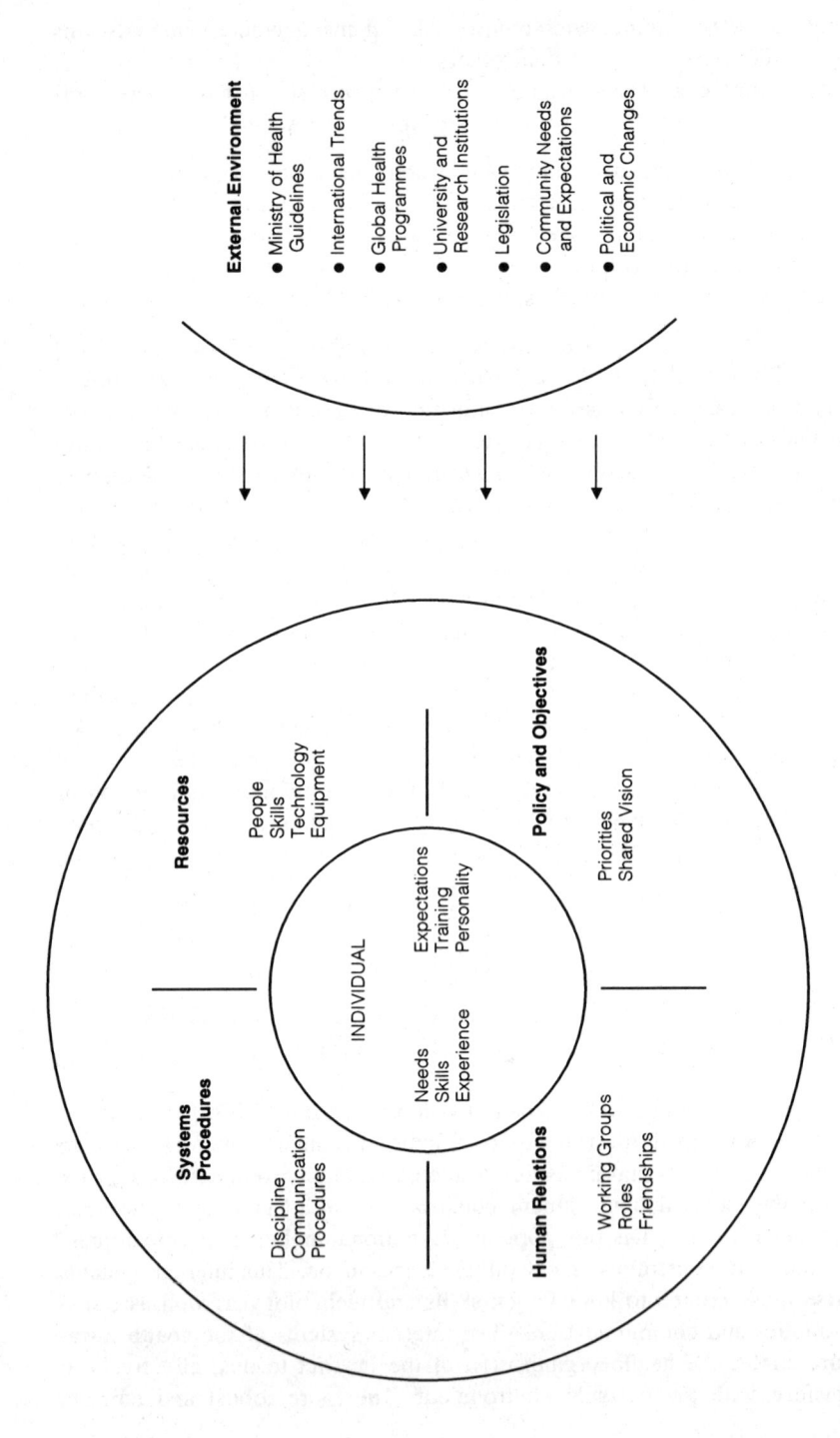

External Environment
- Ministry of Health Guidelines
- International Trends
- Global Health Programmes
- University and Research Institutions
- Legislation
- Community Needs and Expectations
- Political and Economic Changes

Resources
People
Skills
Technology
Equipment

Policy and Objectives
Priorities
Shared Vision

Systems Procedures
Discipline
Communication
Procedures

Human Relations
Working Groups
Roles
Friendships

INDIVIDUAL
Expectations
Training
Personality
Needs
Skills
Experience

Figure 4.1 Health organisation as an internal system within an external environment

the internal system the more effective is the organisation in dealing with the external environment.

The District Health Manager ensures efficient working in three directions:

(1) All the current health functions in the District must run efficiently.
(2) The potential of each individual member of the District Health Team must be realised to the fullest extent.
(3) The District Health Organisation must evolve continuously in response to the changing health needs for example in the 1980s to meet the challenge of PHC, in the 1990s for the post PHC decade and after that for the post Health for All 2000 era. Each task requires a different approach, and yet all these are interrelated.

To tackle any of these tasks, let alone all three together, requires an understanding of the true realities of the health situation in general, and of the District Health Organisation and its capacity for effective performance. However, some realities need to be faced:

(1) The results of the health work are determined *outside* the health organisation – in the health facilities and in the community. It is only here where it will be decided whether the efforts of the District Health Team are producing measurable benefits or are a waste of resources.

   One resource fully under the control of the District Health Management is knowledge of all kinds – from scientific and technical knowledge to social, economic and managerial knowledge. Indeed the function of the District Health Management can be described as converting knowledge into tangible health results.

(2) Results are obtained by exploiting opportunities, for example in epidemiologic or scientific spheres, in community organisation, in programme development or in political change. Rarely are results obtained by solving problems however urgent they may seem at the time. All problem solving does is to eliminate a restriction on the capacity of the organisation to obtain results. One cannot shrug off problems. But to produce results, resources are best directed towards opportunities than to processing. Such a decision for maximisation of opportunities requires sound leadership.

In a District with a large population health events are occurring all the time requiring all kinds of decisions for allocation of resources. But it is the 10 per cent or so of crucial decisions that are likely to produce 90 per cent of the results. Good leadership requires the insight and steadfastness in allocating maximum resources (for example personnel, equipment etc.) to the 10 per cent or so activities crucial for best results. In order to do so, it is necessary for the District Health Manager to see the health activity of the District in its entirety. He must be able to see the resources and the efforts as a whole, and to see their allocation to programmes, to consumers of

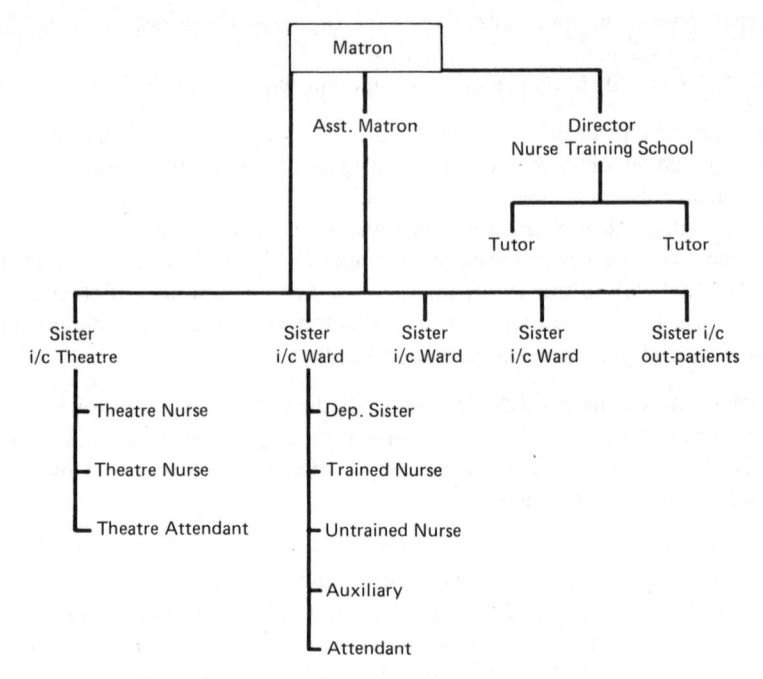

```
                          ┌─────────────┐
                          │   Matron    │
                          └──────┬──────┘
                 ┌───────────────┼───────────────┐
          ┌──────┴──────┐                 ┌──────┴──────┐
          │ Asst. Matron│                 │  Director   │
          └─────────────┘                 │Nurse Training School│
                                          └──────┬──────┘
                                          ┌─────┴─────┐
                                        Tutor       Tutor
```

```
Sister        Sister      Sister      Sister      Sister i/c
i/c Theatre   i/c Ward    i/c Ward    i/c Ward    out-patients

─ Theatre Nurse      ─ Dep. Sister
─ Theatre Nurse      ─ Trained Nurse
─ Theatre Attendant  ─ Untrained Nurse
                     ─ Auxiliary
                     ─ Attendant
```

Figure 4.2 An organisation with relatively few different types of activity

services and their distribution channels. He must be able to see how much effort goes on to problems and how much on to opportunities, and then weigh up alternatives of channels and allocation. Only the overall view of health work in the District as a system can provide a real insight into identifying opportunities for maximum results.

FORMAL AND INFORMAL ORGANISATION

A distinction can be made between a 'formal' and an 'informal' organisation, each of which has advantages in certain situations.

A formal organisation is often described by means of an 'organisation chart' such as those which appear in figures 4.2 and 4.3.

Among the advantages of formal organisation charts are that they:

(1) define broad areas of job responsibility;
(2) provide a basis for writing job descriptions;
(3) indicate channels of communication;
(4) clarify relationships between people;
(5) avoid complications caused by overlapping of functions.

Probably their most useful function is in the thought process the manager must go through in drawing up the chart in the first place. It requires thinking about who actually is in the organisation, what work they do, and whom they relate to.

Figure 4.3 shows the managerial and communications structure of a Primary Health Care organisation. Level B (the Health Centre) teams form the crucial Middle Management level between the local community workers and the District Health Management Team. The position on the chart of the Town or Village Development Committee indicates its important managerial position in relation to Level B staff. These staff are technical advisers to the Town or Village Development Committee. The local community workers come directly under the management of this committee for everything except technical matters like selection, discipline, and so on.

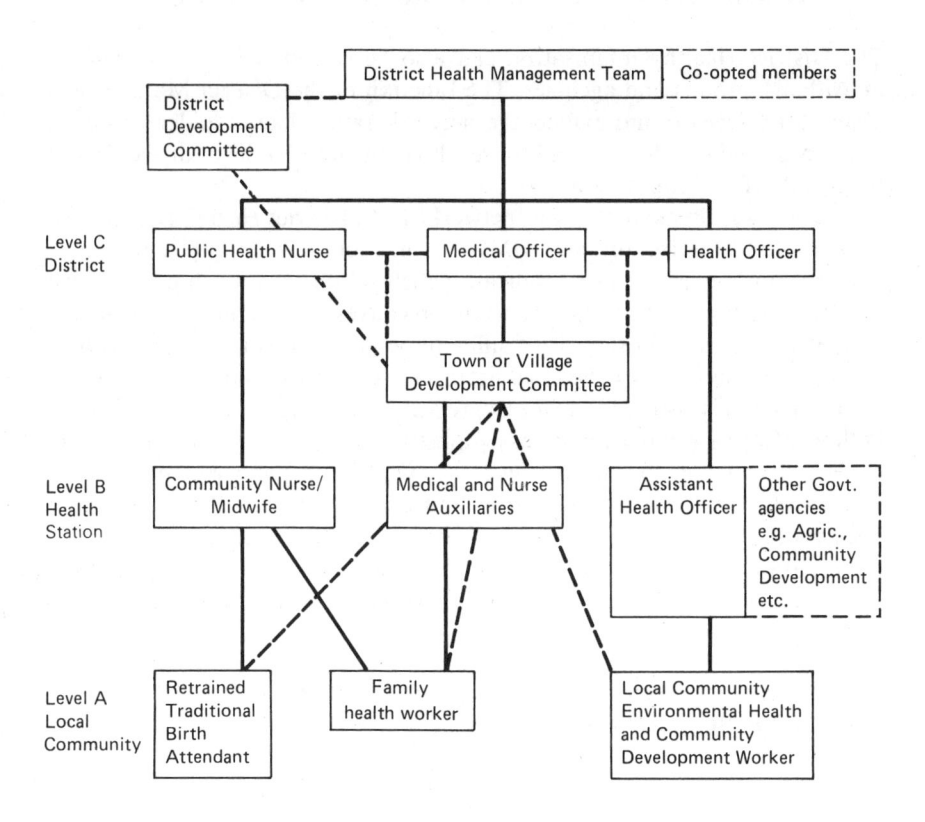

Figure 4.3 Organisational structure of a primary health care system showing managerial and communications structure (adapted from Ministry of Health, Ghana, 1980)

AN ORGANISATION AS A SKILL PYRAMID

It helps to look upon a District Health Organisation as a 'skill pyramid', as in figure 4.4. Such an attitude recognises that there are only a few health workers with particular medical skills and knowledge, and that their function is to diffuse their expertise throughout the community by means of the health organisation. At the same time, there is a great deal of local knowledge and information about the community and its health which needs to come into the organisation and to be used for planning or for the organisation of services. The effective District Medical Officer and his Management Team are concerned to achieve and sustain an organisation in which skill and responsibility are pushed 'down' the organisation as far as possible, and in which information can flow 'up' the organisation as quickly as possible.

AN ORGANISATION AS A NETWORK OF INDIVIDUALS

The District Health Organisation can also be considered as a *network* of individuals, groups and agencies. The function of the District Management Team is to develop and extend the network rather like a spider's web, to encompass and knit together all those who contribute to the improvement of the health of the population.

Figure 4.5 is an example of a 'network' for a District Health Team. It can be a useful exercise for any DMO to map out his or her 'network of relationships'. It is useful to indicate which of the relationships are most important so far as achieving the current objectives is concerned, and what the current state of relationships is to different people. This can then show which relationships need to be developed more so that any necessary improvements can be made. It is also useful to look at relationships in terms of expectations – both what you are expecting other people to do for you and what they expect of the health team. In considering these expectations, and where necessary making changes through discussion and agreement with the people concerned, the District Health Team can do much to direct the attention of the organisation and of the people in it to the matters of most importance, besides thinking seriously about the team's priorities in the most efficient use of time.

AN ORGANISATION AS A SYSTEM OR SERIES OF SYSTEMS FOR GETTING THINGS DONE

We can think of an organisation as a system which is a collection of inputs, processes and outputs. In a health organisation a *simple common integrated system is needed* to bring together a number of separate systems concerned with the movement of drugs, supplies, information, people, referrals and cash.

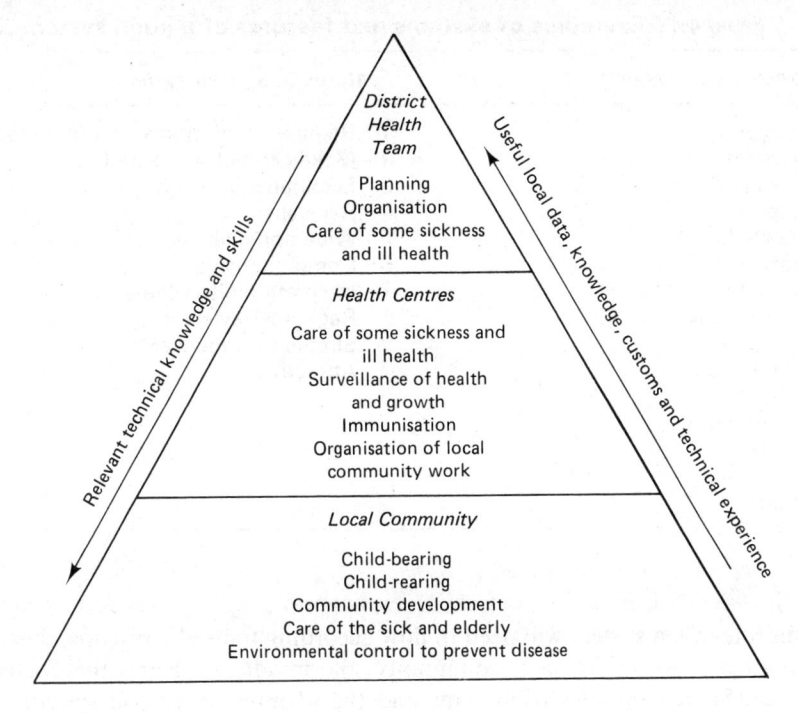

Figure 4.4 An organisation as a skill pyramid

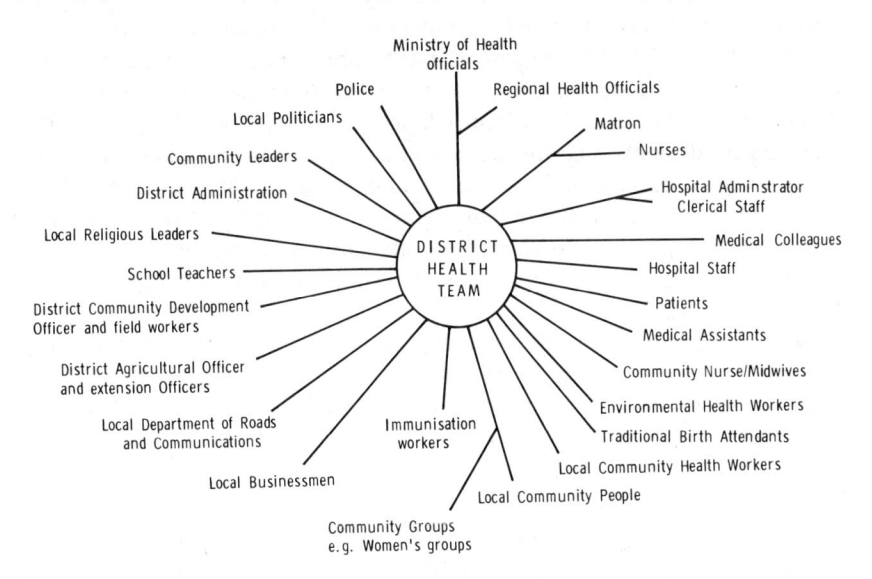

Figure 4.5 The network of relationships for the district health team

Table 4.1 **Examples of systems and features of a good system**

Examples of systems	Features of a good system
Budgeting	1 Regulation of inputs and outputs.
Supplies	2 Makes complexity simple.
Prescribing	3 Eliminates oversight.
Referral	4 Relevant.
Maintenance	5 Wide applicability.
Transport	6 Comprehensive.
Payment	7 Integrated with others
Supervision	8 Regularly reviewed.
Planning	9 Simple to understand.
Information	10 Attractive to use.
Evaluation	
Recruitment	
Training	
Accounting	
Audit	

This integrated system will need to flow according to need daily, weekly and monthly between the local community, the middle level and the District centre. Some examples of systems and the features of a good system are considered in table 4.1.

Two major 'systems' can be distinguished which influence Primary Health Care. These are the 'local traditional system' of decision-making and organisation and the 'imported bureaucratic management' approach.

Many of the ideas of management in the bureaucratic system have evolved in relation to large organisations in industrial and commercial enterprises and public services, particularly in North America and Europe. This approach to management developed alongside rapid economic growth in those countries, and has been a basis for the organisation of government and state services. This kind of bureaucratic management is best exemplified in the large international organisations like the multi-national companies, or in international agencies. For carrying out certain kinds of tasks on a large scale such a system is ideal.

The best example in the health field is probably the systematic eradication of smallpox which was effectively planned, organised and carried out on a global scale with the backing of the World Health Organization. When the local health organisation receives funds from government sources and is part of a National health strategy, then many of the techniques of this type of management will be needed.

Before such a systematic approach came to be developed, however, there were other ways of getting things done. The traditional system still exists in many communities. Although it may appear to differ from place to place there are often common features in this type of organisation in many societies. Features include respect for the views of elders in the community, decision-making only after lengthy discussion by the whole community, formal means of appointing leaders often on a hereditary basis, recognised ways of providing help for members of the community in need, and punishment for those who transgress the community's legal and moral code. In developing Primary Health Care in any community it is folly to ignore the way in which the community organises itself, and this is particularly so in the more traditional outlying rural village communities.

Successful organisation of Primary Health Care therefore requires the ability to understand and work within the two systems, viz. the modern bureaucratic form of management, and the traditional. Diagramatically (see figure 4.6) the two systems can be represented as pyramids: the modern management system as an inverted pyramid with its weight and strength in national and governmental policies, plans, resources and organisation which are translated through the District Health Organisation to create an impact on people living in their rural, peri-urban, and urban communities; the traditional system as a pyramid on its base, with its weight and strength widely distributed in local communities but also having an impact in the life and organisation of the country at District, Provincial and National levels.

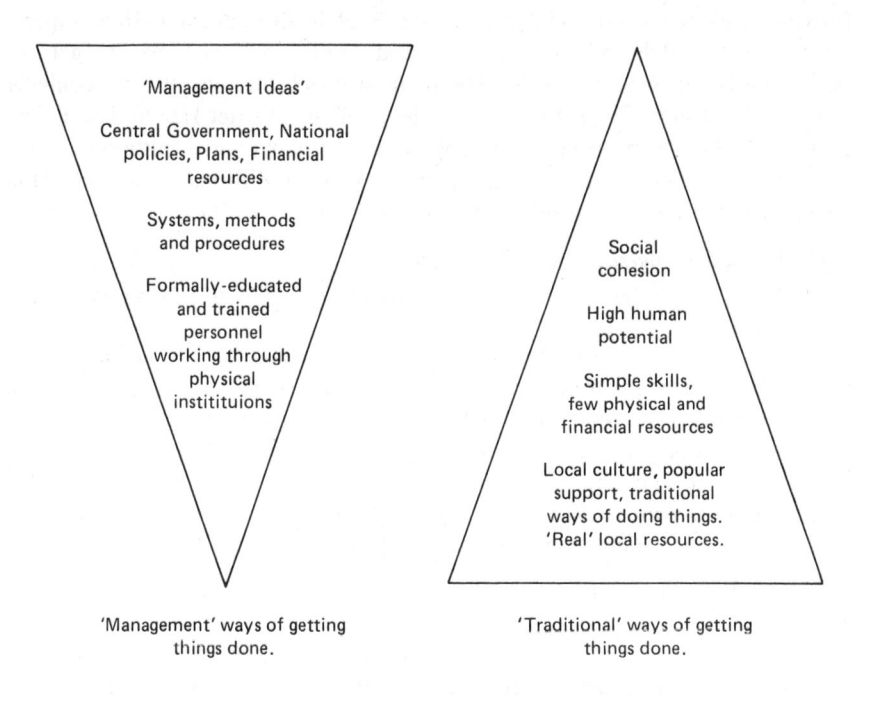

Figure 4.6 'Management' and 'traditional' ways of getting things done

TYPICAL ACTIVITIES	TYPICAL INSTITUTIONS	LEVEL
Interchange of ideas Strategic planning International approach	World Health Organisations International Aid Agencies Multi-National Drug Companies	INTERNATIONAL
National planning and priorities High professional training (Teaching hospitals)	Ministry of Health Institute of Traditional Medicine Council of Chiefs	NATIONAL
Provincial planning Professional training	Provincial/Regional Medical/Environmental Health Officials	PROVINCIAL
Organisations and running of District services Support to locally-based health activities	District Health Team District Hospital Paramount Chieftaincies Renowned traditional healers	LEVEL "C"
Practical and simply organised preventive and curative health services	Health Stations Local Chiefs Local traditional healers	LEVEL "B"
Healthy living as a normal part of village life Basic preventive and first aid services	Local Community Council Local Health Workers Traditional Birth Attendants	LEVEL "A"

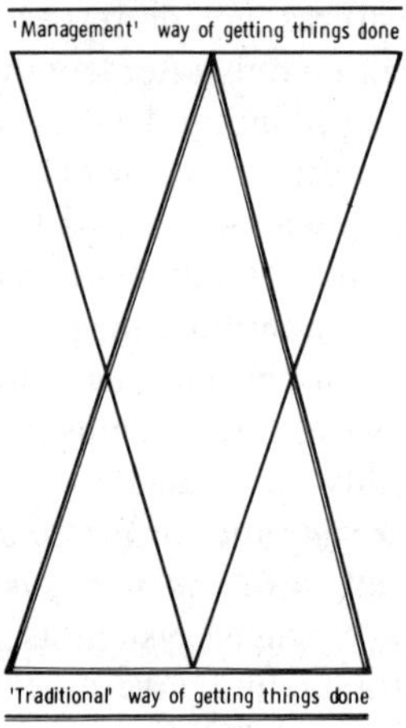

Figure 4.7 Intersection of 'traditional' and 'management' systems

If we next consider a rural health care system in terms of Level A (local community health workers), Level B (Health Centre serving a number of villages), and Level C (District, including hospital services) linked to the Provincial or Regional and National levels of health care and then superimpose our 'models' of 'management' and 'traditional' systems, we get an interesting result (see figure 4.7). The point at which the two systems coincide to the greatest extent is at Level C, the level of the District Health Team. The District Health Team becomes the key point in the over-all health system for integrating the concept of 'management' and 'traditional' systems. This means that the District Health Team should have:

(1) a good *knowledge* of both systems;
(2) a good *understanding* of the strengths and weaknesses of both systems;
(3) the ability to work within both systems so that the strengths of each can be brought together to produce the most effective health system;
(4) the ability to *interpret* to those who are predominantly involved in one system, the contribution which the other system can make towards solving their problems;
(5) the ability to be *creative* in adapting and devising new systems which take into account the advantages of both.

The traditional system is already being influenced by the external system. The challenge is now to enable the traditional system to take from the external system those things which are genuinely useful in a way which does not destroy its own effectiveness, culture and values. In the process the external system may be influenced by what is good in the internal traditional approach.

ORGANISATIONAL CULTURE

All organisations develop cultures of their own. By organisational culture we mean beliefs, symbols (for example uniforms; formal and informal clothing), norms of acceptable behaviour, values, and ways of doing things. Since the tasks carried out and the way they are carried out depend so much on the culture of a particular organisation, it is useful to develop some insight into it. In most post colonial countries there was a legacy of an authoritarian culture wherein decisions were taken centrally by individuals in distant places, and perfect obedience was demanded of those who had to carry them out at the front line. With decentralisation the culture of the District Health Organisation is changing, and may take any of the following well recognised forms, or a mixture of them.

(1) *Power culture*. The person at the leadership position exercises maximum power. For small organisations and with charismatic leaders such an arrangement may be ideal. The lines of communication are short in small organisations, response to crises is quick, and much gets done.
(2) *Role culture*. All employees occupy a set role. Most bureaucracies function along the lines of the role culture with set rules, procedures and a list of general orders. Such a culture is useful for routine tasks, but difficult to change for responding to new challenges. That is why all bureaucracies function in a slow ponderous manner.
(3) *Task culture*. The emphasis is on getting tasks done, and obtaining results. It can respond better to change, and is usually attractive to younger people who are keen and energetic.
(4) *Person culture*. People remain individualistic even though working together in the same organisation, for example consultants, researchers etc.

Most organisations have a mix of cultures depending on their size, how the work is organised (for example by tasks), the environment, and their history.

SOME KEY PRINCIPLES IN AN EFFECTIVE ORGANISATION

Much is written about 'Principles of Organisation'. Table 4.2 lists ten Principles of Organisation. The District Health Team will probably be most concerned with *co-ordination*, making sure that the different parts of the organisation (for example, local community workers, Health Centre staff) do not work in isolation but work together towards a common goal. It must be clear also where authority and responsibility lie for any individual at any one time.

Take the example of a nurse in training who accompanies the epidemic emergency team to gain practical experience as part of her training programme. The nature of her relationship to the team leader, to the nurse on the epidemic emergency team, to her Nurse Tutor, and so on needs to be clear if she is not to be put into the position of having to take contradictory instructions from different people, or of not knowing whom to approach with a particular problem.

Table 4.2 **Some principles of effective organisation**

1 *The principle of co-ordination*
   The purpose of organisation is to facilitate co-ordination; unity of effort.
2 *The span of control*
   No person should supervise more than five, or at the most, six, direct subordinates whose work interlocks.
3 *Definitions and job descriptions*
   The content of each position, both the duties involved, the authority and responsibility contemplated and the relationships with other positions, should be clearly defined in writing and published for all concerned.
4 *The principle of continuity*
   Re-organisation is a continuous process; in every undertaking specific provision needs to be made for it.
5 *The over-all objective*
   Every organisation and every part of the organisation needs to be an expression of the over-all aim of the undertaking. Otherwise it is meaningless and therefore redundant.
6 *Authority*
   In every organised group the supreme authority must rest somewhere. There should be a clear line of authority from the supreme authority to every individual in the group.
7 *Responsibility*
   A superior is always completely responsible for the acts of a junior worker.
8 *Correspondence of responsibility and authority*
   In every position the responsibility and the authority should correspond.
9 *The principle of balance*
   It is essential that the various units of an organisation should be kept in balance.
10 *Specialisation*
   The activities of every member of any large organised group can sometimes usefully be confined to the performance of a single function.

Span of control is also important where there are large numbers of scattered local community health workers. To give them adequate supervision and support may entail grouping them under the supervision of perhaps a community nurse midwife or a medical assistant based at a Rural Health Centre. If the span of control is too large, for example, with very large groups to supervise, people get neglected. If too small, there is wasteful duplication. *Job descriptions* are a well-established way of dealing with the principle of definition, to ensure that the duties, authority and relationships of each post are clearly defined within the organisation. Table 1.13 (The PHW profile) in chapter 1 and table 4.3 below are broad statements of the job requirements for two jobs, a Primary Health worker, and a medical assistant, prepared as guides for those working in the rural health field.

Table 4.3 **Job description of a medical assistant**

Organisational Relationships

Functions

Administration	. Administration of the Health Centre.
	. Supervision and co-ordination of the team of health workers.
	. Supervision of satellite health posts and dispensaries.
	. Production of quarterly reports to District or Regional MOH.
Curative	. Diagnosis and treatment of illness (or referral).
Community Work	. Implement Health and Research Community Programme.
	. Encourage members of community to be involved in health programmes.
	. Convene team members to co-ordinate activities.
	. Participate in village development programmes.
	. Collect and report data – sociological/epidemiological, morbidity/mortality.
	. Give support to VHWs – regularly in their villages – through training activities in the health centre and elsewhere.

Source: *Health Auxiliaries and the Health Team.* Eds Muriel Skeet and Katharine Elliott, p. 156.

These are general descriptions but it is important always to make the job description fit specific situations. A good way of doing this is through discussions aimed at clarifying exactly what a job holder is trying to achieve in his or her work. There are dangers in job descriptions which are too detailed and rigid. Any job is likely to change in the course of time, and different people are likely to do the same job in somewhat different ways depending on the background, interests, experience and so on of those individuals. One way of avoiding rigid job descriptions is to *put an emphasis on the 'outcome' of a job or what* has to be achieved in a person's work, rather than a detailed specification of *how* that work is to be done (see figure 4.8).

The principle of *continuity* is an important one. As conditions change, staff come and go, new ideas are introduced and there will be a need to change and improve the organisation. Major organisational change usually causes upset to the people concerned, but by being in close touch with the organisation and

154 District Health Care

Job Analysis

To analyse and make clear what is involved in doing a particular job

Job Evaluation

To 'evaluate' what a job is worth, i.e. how much a person should be paid

Training

To specify the training a person may need to do a job effectively

JOB DESCRIPTION

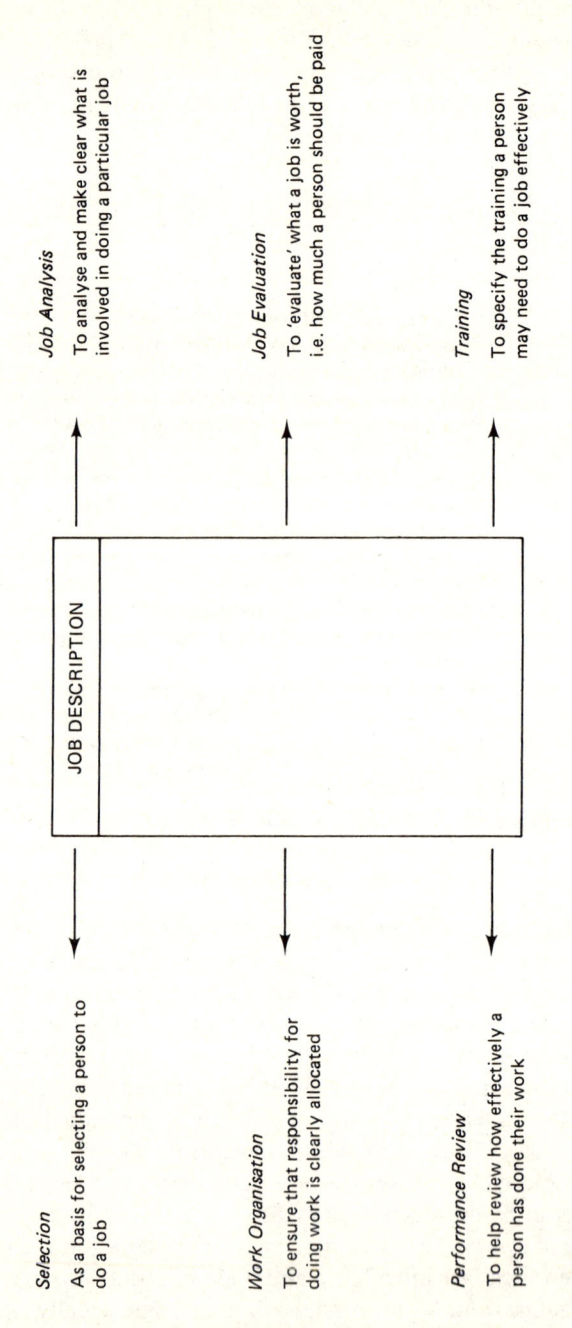

Selection

As a basis for selecting a person to do a job

Work Organisation

To ensure that responsibility for doing work is clearly allocated

Performance Review

To help review how effectively a person has done their work

Figure 4.8 Uses of a job description

anticipating and making changes as they become necessary, senior managers can often avoid the need for more radical changes later.

THE KEY ELEMENTS OF AN EFFECTIVE ORGANISATION

With a little skill and some experience it is not difficult to view an organisation and detect where changes need to be made. The good manager identifies malfunctioning, diagnoses particular problems and prescribes corrective action. But it is then necessary to get the organisation to take the corrective action which may be the most difficult part of the process. As with health, so also with organisations; prevention is often better than cure. The effective manager will constantly be taking action to prevent the organisation getting into major difficulties. There are a number of elements to an effective District Health Organisation, as shown in table 4.4.

Table 4.4 **Four key components of an effective district health organisation**

Analysis of problems
Development of ideas to tackle them (plan)
Administration of supplies, equipment, transport, personnel
Leadership of people

Each of these four topics has to be organised well (see figure 4.9). The process of setting tasks which needs to be developed as a result of problem analysis has already been described.

The planning process then becomes the development of a course of action. It entails the gathering of data, identifying the causes of problems and developing alternative solutions. This needs setting priorities and the development of policies. It also needs resource allocation through budgeting, then programming, formulating strategic plans on how and when to achieve goals, specification of the end results expected (objectives) and targets to establish where the present course of action will lead.

The administrative organisation is simply the arrangement of work to accomplish the objectives effectively. It requires definition of the skills needed for the tasks to be performed in each position in the organisation. Each needs a specification of scope, of responsibility and authority and a defined relationship to others.

People recruited into the organisation need to be provided with a setting in which they can work purposefully and effectively towards the objectives. Finding the right people is clearly crucial. After selection they need to be familiarised with the situation, trained, and then helped to improve their knowledge, attitudes and skills. Their responsibilities and accountability need

156   District Health Care

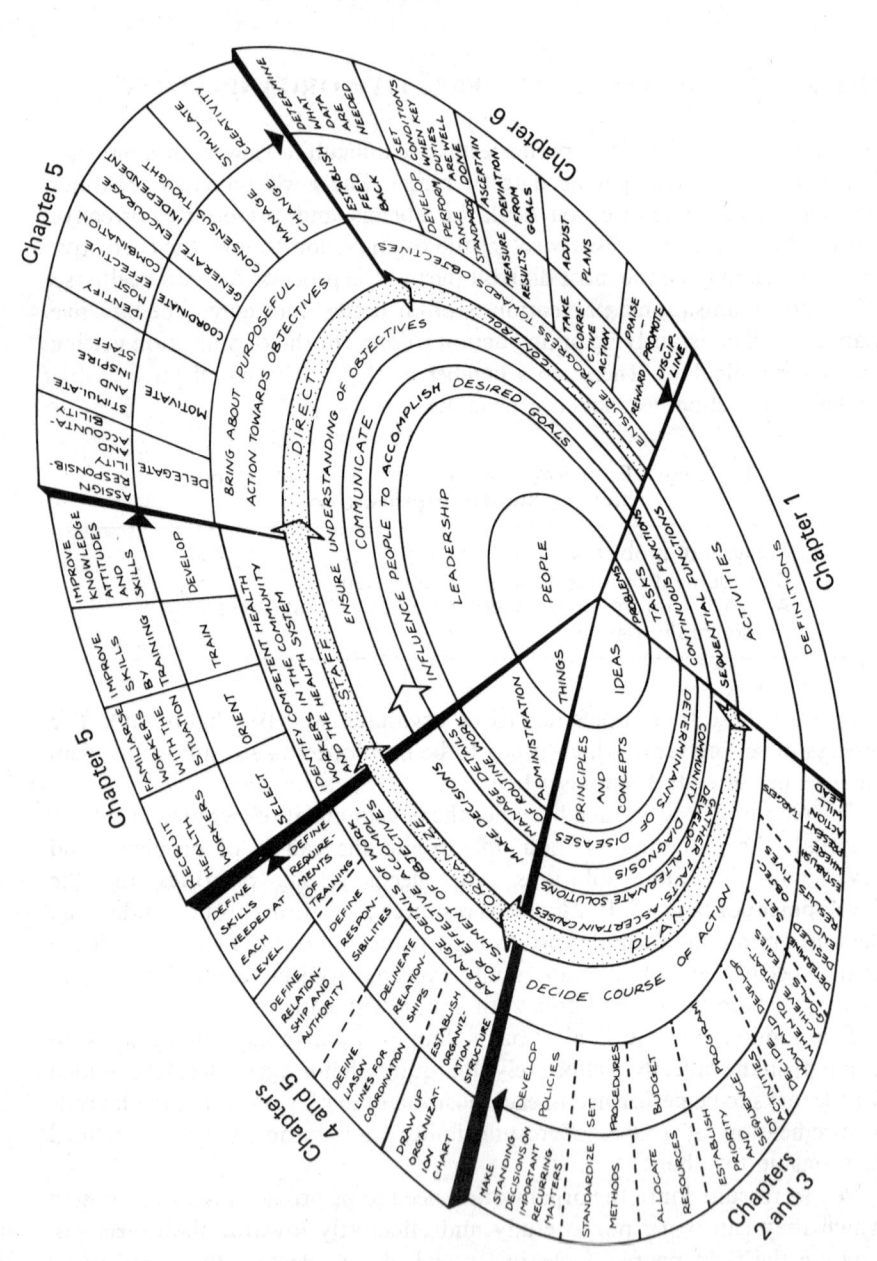

Figure 4.9 Key elements of a district health plan and organisation

careful defining if delegation and co-ordination is to work. Independent thought needs to be fostered and people inspired or encouraged to think for themselves. Differences in opinion need to be managed and conflict resolved. Change needs to be managed too, through stimulating creativity and innovation in achieving goals. Control is needed to make sure that progress is made towards the objectives of the plan. A reporting system is needed so that people know what data are required, and how and when to produce them. Performance standards may be set which indicate what specific conditions exist when key duties are well done. Results can then be measured from both these sources to see any deviation from standards, and corrective action can be taken to adjust plans and standards if needed. Discipline may be necessary at times, but reward, praise and remuneration are just as important as means of control and incentive.

What it means to manage a district health organisation

In order to work efficiently a Rural Health District has to be well organised and run. The District Health Team and the District Medical Officer need to appreciate that their work is to a large extent managerial. Unfortunately in the past doctors and other health workers have not been trained to think of themselves as managers but rather tended to take their teachers, the hospital consultants, as models. So we need to try and answer the question: what does it mean to be a manager? This is shown in table 4.5.

Table 4.5 **What does it mean to be a manager?**

1 Seeing the whole picture (i.e. beyond the hospital, and the entire District as their area of responsibility).
2 Good personal motivation.
3 Recognising contributions which further the aims of the organisation.
4 Effective use of time.
5 Standards of performance.
6 Clear objectives.
7 Planning the future while managing the present.
8 Organising and allocating resources.
9 Decision-making.
10 Delegating, motivating and developing other people.
11 Developing and maintaining systems.
12 Reviewing and evaluating.
13 Being an agent of change.

(1) *Seeing the whole picture*

The problem of many professionals and specialists is that they see things in the light of their own expertise and do not find it easy to appreciate the point of view of others. But the manager must have a broader view. In fact, a manager needs to be a professional 'non-specialist'. All the advice available from specialists is taken, but it must always be put into the context of the whole situation. District Health Managers need to be able to see the whole of the Health District, its needs and problems, as it were, from without rather than from within.

(2) *Good personal motivation*

The attitude of a manager sooner or later affects the whole organisation. If the manager is apathetic, not interested and disillusioned about work, then these attitudes will spread to others, and even keen and enthusiastic workers will stop trying, or look for other ways of channelling their energies. So managers need to be sure that their job is worth doing, and that they are totally committed to it.

(3) *Recognising contributions to the aims of the organisation*

The good manager has a very clear idea of what contribution good management can make to the total success of the work of the Health District. From time to time he will look up from his work and ask 'What is the best contribution I can make to the goals of this Health District?'. At the same time, the manager asks about other workers in the district 'What is the best contribution this person can make?' – both now, through having important work to do, and in the future after appropriate training for making an even more useful contribution.

(4) *Effective use of time*

Good managers know how their time is spent, what is the best use of their time and how to help others use their time well.

(5) *Standards of performance*

Good managers set high standards both for their own work and for the work of others. But for this to be so, a manager must have a clear idea of what such standards are. Standards can range from such things as standards of cleanliness or standards of competence for workers carrying out certain tasks, to standards of coverage of basic health care in the District. Doctors are used to thinking of high standards for clinical work, but the same thinking is needed to determine appropriate standards of managerial and organisational practice throughout the District.

(6) *Clear objectives*

The way to maintain and improve standards is to set realistic objectives which can be achieved within an identifiable period of time. 'Key result areas' are significant aspects of the District's work to which attention should be given to achieve major improvements in health. Managers also need to relate their personal objectives to those of the organisation as a whole.

(7) *Planning the future whilst managing the present*

Planning is an important management tool, but in practice managers have to plan for the future at the same time as they manage the present. It means not only saying 'What needs to be done now?' but also 'What will need to be done in the future and how should we plan for this now?'. In other words, the good manager uses today's problems and opportunities to create a better future.

(8) *Organising and allocating resources*

The main resources are human, material and financial, as well as time, and the manager's task is to see that these are organised in the most effective way to meet the needs of the District. The manager's skill is in making the fullest use of whatever resources are available. Managers who constantly complain about shortages of money expose their own managerial inefficiency.

(9) *Decision-making*

Bad managers either make bad decisions or none at all. Good managers often make relatively few decisions except those of fundamental importance which have a significant effect on the way in which the District is run. Such decisions establish policies and guidelines within which others can work with the minimum of interference. The manager who is constantly having to make decisions about small matters probably needs to give more thought to policy-making, and to the way work and authority are delegated to others in the District. The timing of decisions is often of the utmost importance. Some decisions are more important than others. In all of them, probably the most critical factor is that of judgement. The wise manager takes advice from other people and sources, and often will look for different points of view, but at the end of the day it is his own judgement of the situation and decisions that matters most.

(10) *Delegating, motivating and developing other people*

One definition of a manager is 'someone who is accountable for more work than he is able to do himself', and who is therefore dependent on other people to do some of that work. This means that a manager must be able to delegate work to others, motivate them and win their commitment, and above all encourage, train and develop others to take on greater responsibilities. The skills the manager needs will include *technical* skills, that is, skills in planning and providing health services; *human* skills, that is, skills of working with people, understanding their problems and getting the best from them; and *balancing* skills, that is, skills of being able to see many factors in a complex situation, assess their relative importance and decide an appropriate course of action.

(11) *Developing and maintaining systems*

Much of the work of a Health District is routine and lends itself to being organised in a systematic way. Such things as the regular supplying of drugs and dressings, the organising of clinic sessions and regular training programmes, need to be based on clear and well-understood systems. A manager needs to review regularly such systems and to adapt them in the light of new circumstances which may arise.

A 'systems approach' uses systems wherever they are appropriate, and to link the various systems together to work smoothly. Care may be needed to develop systems with 'fail-safe' mechanisms, so that work will continue to be done in the most difficult circumstances and if need be by people who have often had minimal education and training.

(12) *Reviewing and evaluating*

Managers will need continually to assess how well the work is being done and what difficulties or problems arise. Much of this can be done on an 'exception' basis, for example identifying exceptionally low or high attendances at clinic or health education sessions. Standards of performance give the manager a very clear idea of the standard of work to be expected. Evaluation and review is far more than the application of certain techniques. It is a whole attitude of mind and style of doing things. Good managers recognise a natural propensity for things to go wrong, as well as often to go right! Any situation therefore is reviewed as a matter of course and corrective action is taken when needed.

(13) *Being an agent of change*

No management situation, especially in the development of Primary Health Care systems, is ever static, and changes are constantly having to be made. Small changes may be made easily in the day to day course of work, but big changes need to be thought out in detail and planned carefully. Management of change is one of the key aspects of a manager's job. Although it is important to maintain a general level of stability in the provision of services within the District, there will be very few times when some significant change or another is not taking place.

Managing within the local socio-cultural environment

To be effective, managers must always work within the limits set by the local environment. In many societies this entails taking into account such things as the social structure of rural life, village decision-making processes, respect for elders, obligations within the extended family, established ways of helping individuals in need, accepted legal and moral codes, and norms of punishment. Together such elements help to make up the way a particular society works and in effect constitute a community's own system of management. An effective health manager therefore needs to work both within this 'local' type of system, and also the type of system with which much of this book is concerned. The manager thus becomes a 'bridge' or an 'integrator' bringing together what is good in both systems to produce a workable means of improving and maintaining the health of the community (see figure 4.10).

Various elements of a District Health Organisation exist in most countries. In some these elements come together to form a viable system with a variable degree of effectiveness. In others the District Health System is fragmented with different parts functioning independently and only rarely coming together to influence the health problems. The challenge for the manager of the Health Team is to assemble the various parts of the District Health Organisation into a unified system which can work with efficiency. Some of the principles of creating such a District Organisation have been described in this chapter. Only with an adequate District Organisation does a mechanism exist for the implementation of the health plans. Without such a mechanism the plans remain only on paper. In the next chapter we go on to discuss how plans can be implemented utilising an effective District Health Organisation.

THE DUAL RESPONSIBILITY OF
The District Health Manager

'I must keep in with the people in the system who really matter'
- Bureaucrats
- Officials
- Other managers
- Leaders

'I must keep in with the people in the community who really matter'
- Poor villagers
- Mothers
- Health Workers
- Village Leaders

Techniques
Systems

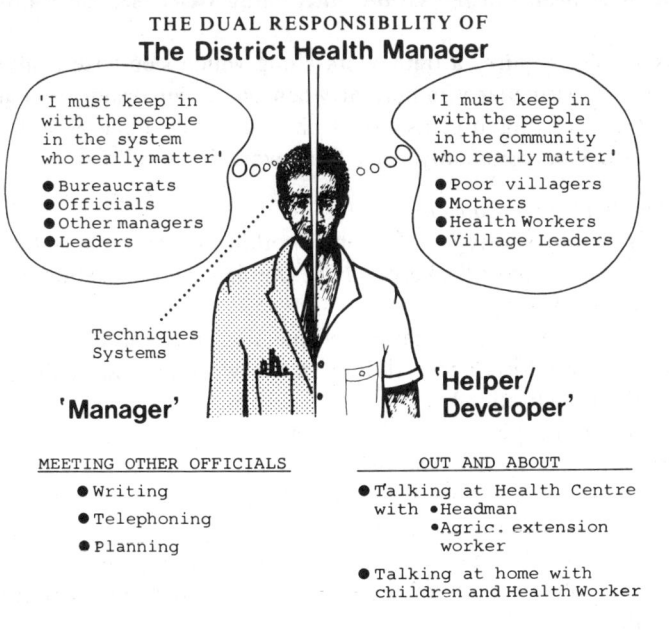

'Manager'

'Helper/ Developer'

MEETING OTHER OFFICIALS
- Writing
- Telephoning
- Planning

OUT AND ABOUT
- Talking at Health Centre with
  - Headman
  - Agric. extension worker
- Talking at home with children and Health Worker

Figure 4.10 The dual responsibility of a district health manager

Table 4.6 **Symptoms of defective organisation**

Motivation and morale are low because:
- decisions appear inconsistent and arbitrary in the absence of defined roles.
- people feel they have little responsibility, opportunity for achievement and recognition because of lack of delegation.
- there is a lack of clarity as to what is expected of people and how their performance is assessed.

Decision-making is delayed and not to the point because:
- necessary information is not transmitted on time.
- decision-makers are overloaded due to insufficient delegation.
- there is no procedure for evaluating similar decisions made in the past.

Conflict breaks out from time to time because:
- there are conflicting goals between projects.
- mechanisms for liaison have not been laid down.
- people involved in operations are not involved in planning.

Rapidly rising costs because:
- there are too many higher level workers.
- there is an excess of procedures and paperwork.

Is the district health organisation functioning well? (see table 4.6)

Any assessment requires a framework along which step-wise analysis can be made. This is particularly important when one is analysing the current situation and considering organisational change in the light of it for meeting future needs. One commonly used framework is as follows:

(1) *People in the organisation*
- What are the types of people working in the organisation?
- What are their backgrounds of training, and education; their needs, fears and hopes?
- How well is the organisation meeting those needs?
- Do the efforts and activities of the workforce meet the purpose of the organisation?
- What changes are needed to answer the above questions?

(2) *Purpose and goals of the organisation*
- What purpose does the organisation serve? What objectives?
- Are the priorities clear?
- How widely shared are the values of the organisation?
- Do different parts of the organisation understand how they fit into the whole?
- How well is the organisation doing what they say they wish to do?

(3) *Structures and systems*
   - Do the structure and systems help the organisation work efficiently and achieve?
   - Are the staff roles clear? Do they make sense as a whole?
   - How well do the systems permit effective day-to-day management? (Systems are the nerves of an organisation. They communicate signals and alert decision-makers to things going wrong, and enable them to plan ahead and co-ordinate.)

(4) *Relationships*
   - How do the formal and informal relationships between groups and individuals help or hinder the work?
   - How do the relationships reflect the purpose of the organisation?
   - Where are relationships a problem, and why?
   - How could good relationships be built upon, and poor ones tackled?
   - Is there enough openness and willingness to confront areas of difficulty?

(5) *Resources*
   - What resources – human, material, and financial are available?
   - Are they being used effectively?
   - What resources could be used to better effect?
   - What other resources are needed?
   - How can they be acquired?
   - Would more resources really help?

(6) *External influences*
   - How well is the organisation responding to new legislation governing major aspects of health? What mechanisms exist for doing so?
   - Are community needs in health development being assessed on a regular basis?
   - What economic factors are affecting the working of the organisation?

Organisational change (see figure 4.11)

After a detailed analysis of the different components of the District Health Organisation and the external influences on it, the next step is to consider 'Is change needed and if so what change is most appropriate?' Organisational change can occur in a variety of ways. It may come about as a result of changes in the goals, in the technology being employed, in the membership of the organisation, or a mixture of all these. Sometimes the emphasis is on changes in personnel occupying key positions; sometimes in the grouping of individuals; and at times through changes in structures and systems.

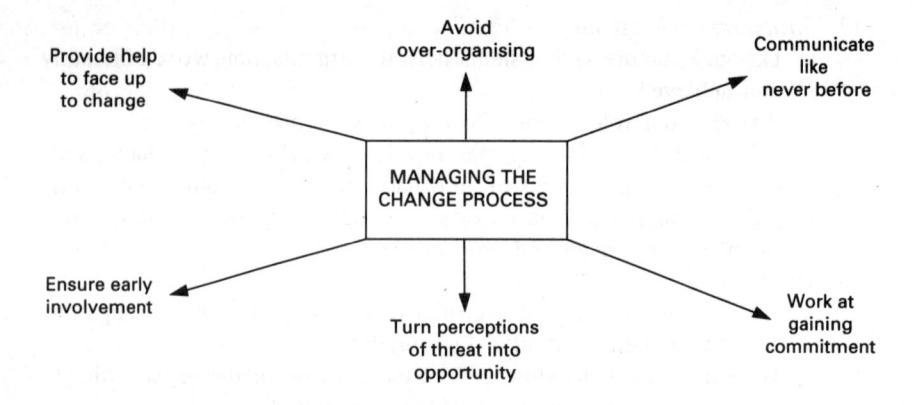

```
                    Avoid
                over-organising
   Provide help                      Communicate
   to face up                            like
   to change                         never before
                 ┌──────────────┐
                 │ MANAGING THE │
                 │CHANGE PROCESS│
                 └──────────────┘
   Ensure early                          Work at
   involvement                           gaining
                Turn perceptions       commitment
                 of threat into
                   opportunity
```

Figure 4.11 Six key activities for managing change

Evolutionary change

All organisations change over time in response to changes in the external influences, or within the organisation. Social currents and changes will bring about changes in community needs. Demographic trends may identify new needs (for example the 'greying' of the population), or a global drive on a previously unrecognised health issue may occur (for example the Safe Motherhood Initiative). New outbreaks and epidemics (for example HIV infection; dengue epidemic), changes in life styles (for example substance abuse), technological change or new discoveries in medical science can all trigger changes in the District Health Organisation. Government policies change from one year to another, or health budgets may alter from year to year and require appropriate action. Within the organisation change happens on account of staff movements, new relationships (or conflicts) developing, new technologies, and so on. Evolutionary change is slow and generally acceptable. Sudden change is threatening and painful.

Resistance to sudden change

Resistance to sudden change is a common reaction. It becomes more problematic when:

(1) The purpose of the change has not been made clear.
(2) Persons affected by the change process are not involved in its planning.
(3) Change is based on personal reasons.
(4) The 'cost' is too high or the reward 'inadequate'.
(5) There is anxiety over job security.
(6) People see vested interests behind the change.
(7) There is lack of respect or trust in its initiator.

When faced with symptoms of malfunctioning in the organisation or reluctance to change, the need for the District Health Manager is to consider all the individual components of the health organisation in turn for introducing change as follows:

(1) *People*

Is it possible to find ways of reconciling individual needs and the goals of the organisation? If there has to be conflict, how can it be rendered positive and constructive?

How can power and leadership be distributed within the organisation?

(2) *Purpose*

How can clearly defined goals be maintained and communicated well so that people remain committed amidst complexity and change?

(3) *Structure and systems*

How can structures and systems be continually renovated to prevent them from going stale?

How can the organisation best adapt to external forces of change?

(4) *Relationships*

How can individuals and groups be helped and motivated to work together to achieve organisational goals in a changing environment?

A strategy is needed to set the direction of change. Within the organisation it is necessary to have everyone committed to a set of values, a strategy and action. Suggestions for ways of introducing organisational change are detailed below, and for changes in individual units and departments on page 215.

Developing a shared vision (see also page 89)

Developing the organisational vision requires key people (doctors, nurses, administrative staff, District health administrators, community representatives etc.) to pool their views with the aim of making planning more imaginative and yet relevant. In the process they become committed. The values, aspirations and constraints are sought from the key people through a process of understanding and compromise which are integrated into the change process. The range of methods to achieve this include:

(1) *Vision workshops*   In these the participants imagine themselves into the future and paint the scenario of their work.

(2) *Brainstorming*	Relevant ideas are gathered and then organised under appropriate headings. SWOT (strengths, weaknesses, opportunities, threats) is a form of brainstorming. A multisectoral SWOT analysis can outline the effects of change on individual units of the District health service. This is also called Domainal analysis (see page 219).
(3) *Buzz groups*	Smaller groups as parts of a plenary group discuss one tightly focused section of the topic.
(4) *Chaired meetings*	Comments are invited, clarification sought and given, and points of agreement made.
(5) *Briefing papers*	One person prepares a paper providing situational analysis. The paper is then used as a starting point for further thought by participants at a meeting.
(6) *Facilitated discussion*	A neutral outsider is invited to help develop the discussion.

The purpose of the activity is shown in the following diagram.

Outline the present → Outline the vision → Identify the gap

Having necessary structures like steering groups, task forces, conferences and assemblies is helpful, but change cannot be managed without leadership. Resistance to change is lessened when:

(1) Those affected by the change are involved in its planning.
(2) Accurate and complete information is provided.
(3) People are given a chance to air their feelings.
(4) Only essential changes are made.
(5) Adequate motivation is provided (for example reward, opportunity for advancement etc.)
(6) People know the reasons for change and the overall goals.
(7) There is a climate of trust.

The levers of change

A skilful manager is continually monitoring the interface between the organisation and the environment. Active management involves constantly

reviewing what is happening every day, week, and month. The review process can be both formal and informal. Change in the functions of an organisation can happen by working the following levers according to requirement:
(1) *People*. One can consider changing the skills, capacities, and attitudes of key individuals rather than changing the people, or bringing in new people. In District health work skills in handling people become increasingly important as the individual manager moves up the promotional ladder. At the lower level technical skills are important. At the middle management level interpersonal skills become increasingly useful. Finally, at the top level conceptual skills are the ones most needed, in addition to those of handling people and professional competence.
(2) *Tasks and technology*. A change of task or its definition can change many things. The introduction of new technology for the performance of tasks may also have a similar effect. But more importantly, the ability to see the proper role and focus of the organisation and being able to communicate that vision is a powerful lever of change.
(3) *Structures and systems*. Structural adaptation is essential to respond to changing needs. By themselves structural changes do not achieve much. People go on performing in the same manner as before unless their task is also modified, the systems adapted accordingly, and they themselves trained for the new function.

FURTHER READING

Handy C. B. *Understanding Organisations*. 3rd edition. Penguin Books Ltd., Harmondsworth, 1985.
Jedlicka A. *Organisational Change and the Third World*. Praeger, London, 1987.
Morgan G. *Images of Organisation*. Sage publications, London, 1986.
World Health Organization. *Intermediate Level Support for Primary Health Care, a Framework for Analysis and Action*, (Offset SHS/82.2). WHO, Geneva, 1982.

5 Practical Management: Putting Plans into Action

Planning should lead to action for achieving results intended in the plan. Managers who have gone through the demanding and difficult exercise of producing a health plan and strategy for their District, develop a strong commitment to putting the plan into practice, but they need all the practical skills and experience available to bring the plan to fruition. However, there are often difficulties. Practical skills are best learnt by doing, and most managers say that they learn how to manage by actually managing. Much of what is called 'management' is common sense and can be recognised as such. This chapter is concerned with what it means to be a manager of a District Health Team concerned with putting a health plan into action. Some techniques will be described which can help managers to organise their own work more efficiently, leading on to a range of issues concerned with managing people, which is at the heart of a manager's work. There is a section concerned with 'logistics' – some of the basic systems which need to work smoothly in a health district; and finally there are practical examples of effective management at three levels – the community, the Health Centre and the District.

The health activities of a District occur by means of a variety of service points. These are the District Hospital, voluntary agency hospitals, health centres, several sub-centres, health posts, mobile outreach services, community activities, and so on. The total number of workers may be more than a hundred. The District Health Manager may not know how they spend their time, how effective their work is, and how changes may be introduced for greater efficiency.

All the staff work as several teams, for example those working in hospitals, health centres, sub-centres and so on. Within a given team there are subgroups, for example looking after in-patients and those helping in out-patients. To ensure clear relationships between the various subgroups in a hospital or health centre there has to be a management structure based on departments. A management team made up of heads of various departments meet regularly to discuss matters affecting the hospital or health centre. A hospital or health centre director provides the overall leadership and chairs such meetings.

It is in such meetings that the staff working in a health facility can identify the aim of their work; for example, whether it is looking after only those who visit the facility, or serving the entire District as part of PHC. How staff see the purpose of the health facility will determine how they do their work.

The objectives described in the District Health Plan and the mission statement provide the driving force and help to navigate the District organisation. The leadership and team managers act as helmsmen to steer the course. In order to progress smoothly towards the goals in the mission statement the helmsmen set up systems and procedures which will make the most effective use of the available resources like staff, material, finance and so on. All these aspects of the work of a District function in balance with each other and influence each other. In turn, they are influenced by external factors which may be called the environment.

Such a framework is the essential backbone of health management at the District level. It consists of the following:

Mission statement and plan objectives This is developed by the leadership of the District Health Management as a process of consensus with team leaders and other workers. The mission statement is for continuing evaluation and discussion at agreed intervals to consider changes. Plan objectives are for the duration of the plan.

Teams and team leaders The health manpower of the District is organised into various teams, each with a team leader to address the objectives as they apply to the team. Within teams people work in various relationships and roles (both formal and informal) with each other and with other teams. The quality of these relationships determine to a great extent the efficiency with which work gets done. Some of the teams are part of vertical programmes with their own chain of commands, for example Tuberculosis and Leprosy control, Family Planning and so on.

Resources and strategies Teams have resources allocated to them from time to time during the financial year, which are deployed along the lines of agreed strategies.

Systems and procedures Different teams work along a variety of systems and procedures which have evolved over time, and are continuing to evolve. There are therefore procedures for dealing with out-patients and in-patients, filter clinics, MCH work, surgical and other forms of treatment, community outreach work, and so on.

Environment — Every enterprise must interact with external influences, collectively called the environment. For the District Health Organisation the environment is the community, its life-style, ecology and resources; the Ministry of Health with its directives and guidelines; the global health initiatives promoted by international bodies like WHO and UNICEF, the NGOs and other bodies working in the District.

It will be obvious that the District Health Organisation is like any other organisation. It has a structure (described as the health infrastructure and comprising Levels A, B and C), people and systems. To take an analogy from the human body the structure is the skeleton; the people are the flesh and blood, and the systems are the nerves or the channels of communication. The systems (like the nerves in our bodies) are to communicate signals to and from the centres. They alert the centres if anything is going wrong. The centre can then think and plan ahead.

Under the general heading of systems are included the following:

(1) *Operating systems* which are concerned with the daily functioning of the organisation like, for example, the inputs, the processes and the outcomes.
(2) *Maintenance systems* which are the linking mechanisms between different sections (preventive and curative; in-patients and out-patients; medical, surgical and obstetrics, and so on); different levels (Levels A, B and C); different sections (transport; logistics; finance; records, and so on). There are also the reward and control systems for the staff.
(3) *Adaptive systems* which are concerned with responding to the current and future needs of the communities served, and with deciding policies.
(4) *Information systems* serve the above three as well as the centre.

It is worth bearing in mind that the systems of an organisation are there to serve all those who work in it. They are put together by humans, and will be operated by humans. Hence a great deal of behavioural science comes into their smooth functioning. Management logic about bringing the work force and resources together to achieve defined objectives must work hand in hand with behaviour science and group psychology if the systems are going to work efficiently.

Planning, organisation, co-ordination and control are usually described as the main functions of managers. In reality, they have very little bearing on the daily routines of managers. From the day-to-day point of view the main activities of managers can be grouped into three areas – interpersonal, informational and decisional.

(1) *Interpersonal role*. This demands three types of functions:
  (a) Figurehead – ceremonial roles like making speeches, presentations, showing visitors and dignitaries around.
  (b) Leader – hiring, training and motivating employees.
  (c) Liaison – a network of relationships within and outside the health organisation mainly for building up an information gathering system.
(2) *Informational role*. The processing of information is a key part of the manager's job. To a large extent this involves communication. In this role the manager is expected to:
  (a) Keep tabs on what is going on.
  (b) Transmit and disseminate essential information to subordinates.
  (c) Be the spokesperson (voice) of the organisation or unit.
(3) *Decisional role*. Making decisions on an on-going basis to improve the efficiency and well-being of the organisation or unit. As a decision-maker the manager acts as:
  (a) Entrepreneur – for example public relations; dealing with cash flow; re-organising weak departments and units.
  (b) Handler of conflicts and disturbance – for example failure of key supplies to arrive on time; periodic outbreaks of disease and scares in the community and so on.
  (c) Resource allocator – deploying to the best advantage the physical and human resources as well as time.
  (d) Negotiator – of contracts; objectives and targets; quality standards, and so on.

MANAGEMENT BY OBJECTIVES

Management by objectives aims to clarify the overall goals of the District Health activity into its component parts. Thereby the responsibility for achieving those goals is evenly and reasonably distributed round the District Health Team. Clearly spelt out objectives identify what each unit manager's team is expected to achieve. The objectives are derived from the District Health Plan (or the mission statement), and unit managers are encouraged to measure their performance against the targets of the plan.

The District Health Manager normally uses three approaches for deriving the objectives of the various teams viz.

(a) Technical and other *information* in deciding strategies.

(b) *Negotiations* with the unit managers and other members of the health team concerned.
(c) A *learning* process of adapting to experience and change.

Performance indicators are also agreed at the same time as the objectives. They are best set according to what is considered important (*Key Result Areas*), and achievable. There is always a tendency to set unrealistic indicators.

Management by objectives seeks to integrate two things:

(1) the achievement of the aims and purposes of the District Plan (as expressed in the Primary Health Care Plan), with
(2) the need for 'managers' (for example, the District Medical Officer, heads of Health Centres, dispensaries, and so on) to contribute to the aims and objectives of the District Health Plan and to develop skills in their own spheres of work.

It is a demanding and rewarding style of managing a Health District.

When a worthwhile system of 'Management by Objectives' is operating, there is a continuous process of:

(1) critically reviewing and even restating the *long-term and short-term health plans* of the District;
(2) clarifying with each 'manager' his or her *Key Result Areas*, which form part of the District Health Plan;
(3) agreeing with each 'manager' *targets* and *action plans* based on an analysis of the problems and resources available to meet them within each Key Result Area;
(4) providing the right working atmosphere in which managers can achieve their Action Plans; reconciling their objectives with those of others; organising their work; and providing training and support where necessary;
(5) giving *control information* in a form which encourages greater efficiency and better and quicker decision-making;
(6) carrying out *regular reviews* to assess progress made and to develop action plans.

'Management by Objectives' is an important supplement to the *planning* activity in a District since it provides a means of translating plans into action by individual managers. Its use can be illustrated by taking an activity likely to feature in many District Health Plans, for example, that of improving potable water supply. This is likely to be one of the Key Result Areas of the manager responsible, say the District Environmental Health and Water Engineer. In practice, managers are not able to deal at any one time with more than a limited number (four or five, and certainly no more than six) Key Result Areas to achieve real results. Key Result Areas emphasise the importance of *results*

Practical management

which are observable, and of measurable improvements within a specified period of time. The Environmental Health and Water Engineer therefore identifies 'improving the supply and use of safe water' as a Key Result Area in his job. Next, he or she lists the main *problems* in the District which affect the supply and use of safe water to the population, for example,

- seasonal rainfall (six months drought each year);
- water contaminated by refuse, excreta and animals;
- poor wells – uncovered, dirty surrounds, dirty buckets;
- shortage of wells;
- storage containers in villages become contaminated;
- difficulties in transport of water;
- river water also used for drinking, washing, swimming;
- villagers do not understand principles of water hygiene;
- non-availability of pit-privees, and improper use of those available.

He or she then considers what can be done about each problem, taking into account what resources are available (for example, money, health workers, support and enthusiasm of villagers) and decides on *targets*. The targets need to be *specific* rather than general, and *realistic*, that is, things that in his or her judgement can be achieved; they should be *measurable* and achievable within a specified *time period*. Thus his or her targets may include such things as:

Within the next year:

(1) to increase the number of usable wells in the District by 25 per cent;
(2) to make contact with six village development committees and agree with them a programme of maintenance of, for example, wells, walls, covers, hoists and buckets to an acceptable standard;
(3) to bring up to date health education programmes on water use;
(4) to extend the 'water' health education programme to market places and public wells in the District;
(5) to contact riverside villages and agree programmes for better use of available water;
(6) to finalise above plans with Public Works Department and prepare budget estimates;
(7) to increase the number of sanitarians in the District from 20 to 25;
(8) to provide basic training for newly-appointed sanitarians and refresher training for all others.

With the help of such targets the Water Engineer has a very clear idea of what is to be achieved during the next year, and can organise his work accordingly. This can be done by listing against each of the targets the precise action to be taken, and who will take it. In this way he produces objectives for each of his staff. In the course of the year the Water Engineer will need to

Table 5.1 Example of method of setting out kind of information needed for Environmental Health and Water Engineer in district

NAME: .... JOB TITLE: Environmental Health Officer LOCATION: .... Health District DATE: March 1993

Key result area	Problems	Resources available to meet problems	Targets	Action plans	Means of control/review
Improve the supply and use of safe water	*Water sources* . Seasonal rainfall . Poor wells . Shortage of wells	1. *Finance* Annual budget +10% special allocation	. Increase number of useable wells by 25%	. Agree location . Allocate layout . Supervise progress	Monthly return on number of wells in use
	Distribution and supply . Poor storage . Transport difficulties	2. *Staff* Environmental Health Workers Other Health Workers	. Start programme with six village health committees	. Discuss with District Medical Officer . Draw up outline programme . Arrange meetings	Reports from village health committees
	Human waste disposal . Mixed use of rivers . Ignorance of water hygiene	3. *Other support* Villagers Schools Market Superintendents	. Update Health Education Programme	. Discuss with Provincial Water Engineer . Get help of Training School	Completed programme by... ....
	. Shortage of latrines . Contaminated wells	4. Public Works Department	. Extend Health Education Programme to markets and wells	. Contact market superintendents and headmen . Train health educators	Reports from market superintendents Inspection of wells

. Agree programme with riverside villages	. Inspect rivers . Report to headmen . Agree programme	Reports from riverside headmen
. Finalise plans with Public Works Dept. for drilling wells	. Agree programme and budget with Public Works Department . Monitor progress	Monthly reports from Public Works Department
. Increase number of sanitarians by 25%	. Agree establishment and budget . Recruit from schools and villages	Numbers in post
. Provide basic and refresher training	. Update training programme . Allocate trainers . Fix training dates	Numbers of trained 'graduates'

check regularly to see what progress is being made to achieve these targets and will need to visit widely within the District to receive *control information* in the form of regular reports from sanitarians, Health Centres, and so on. This feedback of information is important as it enables objectives to be revised where necessary and updated for succeeding years.

Table 5.1 shows a way of setting out the kind of information needed for this type of approach, using the example of the Environmental Health and Water Engineer just discussed.

To be successful a manager should agree objectives in conjunction with other people. Thus in the above example, the District Water Engineer would agree his broad objectives and targets with other members of the District Health Team, thereby benefiting from their ideas, as well as ensuring that his objectives are compatible with theirs. Such an exercise also helps to strengthen the commitment of all concerned. Similarly, by discussing and agreeing targets with the sanitarians, health workers, and others who work for him, he can explain why certain work has to be done, how it contributes to the improvement of health, and thus provide an opportunity for his staff to contribute ideas and share in the planning and running of environmental health work in the District.

As a technique, management by objectives becomes really effective where it is linked to other systems in the District. It is a natural way of translating ideas from the *planning* system into the action system; where objectives are carefully costed it becomes part of the *financial control* system; it is a basis for the District's *Work Allocation* system; it provides a base for realistic policies for *recruitment, training and development* of staff in the District. In some organisations its use has been taken to extremes, but used sensibly it is a technique which can help managers to tackle the day to day work and the future development of the District in a systematic way.

PARTICIPATORY MANAGEMENT

Participatory management is defined as a style of management which creates opportunities for the work force to influence their work and its context. Participation can be formal where representatives of the employees share in decision-making, or informal through job enrichment or teamwork.

A number of strategies have been evolved in industry and commerce for improving the level of participation. Several of the following can be easily adapted for District health work:

(1) *Creating self-regulating work groups*
    Members of the work group take collective responsibility for services provided (for example MCH coverage; Family Planning services etc.). Members of the community may be co-opted into the work group in

Table 5.2 **Factors enhancing the management/staff relationship**

1. A *sense of caring* and concern for the individual.
2. *The environment.* Large organisations, by their very nature tend to be impersonal. Decentralisation of function is needed to generate a feeling of personal involvement amongst the workers.
3. A *family atmosphere* where people are able to relate to each other irrespective of their positions.
4. *Employment security* generated by policies which minimise threats to jobs.
5. *Promotion policies* whereby talent is recognised and further developed for promotion from within the organisation.
6. *Fair levels* of pay and benefits.
7. *Listening* to staff problems and grievances.
8. Managers who have been trained to value a *participative type of management*

cases where community participation is important (for example maternity care).

(2) *Job enrichment of employees*

Providing better facilities for work, opportunities for training, and discretion in work style are some of the approaches tried.

(3) *Project groups*

Groups are set up to handle different projects (for example water and sanitation; HIV infection etc.). Members of the groups may have different backgrounds of training and may even come from different levels of the organisation. Sometimes the project may be a problem solving exercise related to one or more projects.

(4) *Quality circles*

Groups may be formed to address questions of quality of care provided at different service points, and their audit.

(5) *Staff suggestion schemes*

Those working at the front-line often have useful suggestions for improving the quality of work. A scheme which provides incentives for workers to put forth suggestions can increase the sense of involvement provided the management is seen to be acting on them.

Table 5.2 lists factors proven to enhance management/staff relations. The importance of retaining good workers has long been recognised in industry and commerce, and is now being increasingly recognised in service industries including health. But the old patterns still persist. Depending on the value systems of the top management in every organisation there is a shifting balance between concern for people and concern for output. A management grid can be constructed as shown on page 178.

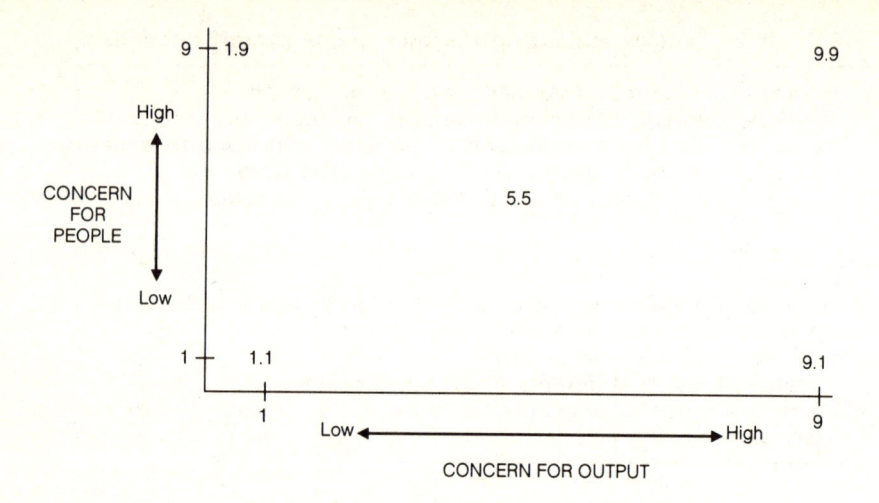

```
         9 ┬ 1.9                                    9.9
    High
     ▲
CONCERN
  FOR                          5.5
 PEOPLE
     ▼
    Low
         1 ┼  1.1                                    9.1
           ┼─────────────────────────────────────────┼
           1                                         9
             Low ◄─────────────────────► High
                    CONCERN FOR OUTPUT
```

9.1 = People are used as just instruments of production.
9.9 = Creativity of the workers is utilised to the full.
1.1 = Little concern for people or production. Workers are left to fend for themselves.
1.9 = Workers' loyalty is bought at the cost of production.

STANDARDS

The question of standards is central to any consideration of Management by Objectives and indeed to the work of a manager. Whether consciously or not, managers are continually making judgements about standards in their day to day work. Such statements as 'X' has done a good job, 'Supplies have been arriving late during the past month', 'Too few patients are attending this clinic', 'We are making good progress in digging new wells', each refer to a standard of some kind. There is a standard of work which is expected of 'X'; a standard of punctuality expected for receiving supplies; some idea (or standard) of the number of patients who ought to attend a clinic; some measure or target for digging new wells in the locality. Managers who set high standards for themselves and constantly work to improve those standards, are in the best position to require and obtain high standards of work from others, provided that those standards are realistic. What is more important is for managers to set realistic standards agreed and accepted by those whose duty it is to implement them. Standards should as far as possible be measurable with regard to:

    Quantity   (for example, the number of patients to be seen in a clinic session)

Quality   (for example, the quality of care provided and hence the need for written procedures for routine treatment and emergencies)
Cost   (for example, average drug cost per patient for common illnesses)
Time   (for example, time taken per patient consultation)

Such standards need to be taken into account in assessing how well a health facility is run. Other standards for a health facility might include:

Access   For example, 80 per cent of the local population have access to services;
                   75 per cent of expectant mothers receive antenatal care (next year we aim to raise this to 85 per cent);
                   no patient has to wait more than one hour to be seen, and 70 per cent are seen within 30 minutes.
History-taking   A history is taken of every patient, which answers at least three of the questions in the Medical Assistant's Manual.
Examination   Every patient is examined before a diagnosis is made.

The following are Effectiveness Standards for Primary Health Care which may be adopted by a District Health Team:

(1) Nothing is done at a higher level of health provision (for example, District hospital) which can be done equally well at a lower level (for example, Health Centre or village).
(2) Until such time as basic health care needs are met a significant proportion (say 50 per cent) of the District's total budget will be spent at Levels A and B.
(3) A significant proportion of time (20–50 per cent) of those with skills at Levels B and C will be spent in training others.
(4) Services provided at Level A, and a proportion (up to 30 per cent) at Level B, will be provided in response to and with the support of villagers, usually expressed through a Village Health Committee.
(5) No major item of spending or change of plan will be incurred by one function (Medical, Nursing, Environmental Health) without the positive agreement of all the District Health Team.

Obviously these are only examples of standards which might be acceptable, and there are limits to having detailed standards for every conceivable situation. What is needed is for health workers and managers to have a clear idea of the standard of work expected. Standards of work are often low because it has never been made clear what level of performance is expected.

In summary, where managers think through in a systematic way as to what objectives and targets they can achieve in future, they are more likely to achieve them; and where standards of performance and work are made clear to workers involved, there is a greater likelihood of improved standards being achieved.

The purpose in establishing standards of work performance is to achieve continual improvements in the service bearing in mind the 4 E's viz. Effectiveness; Efficiency; Equity and Empathy. The practical targets to aim for are: (a) ever-better quality of care; (b) ever-lower costs; (c) ever-increasing flexibility; and (d) ever-quicker response.

Indicators for measuring coverage have been described in chapter 2 (page 57). Similar indicators may be selected for establishing standards of care in other aspects of the health service, for example, curative care, selection of high risk pregnancy and referral, treating the five major killer diseases of children, and so on. Indicators of performance standard are defined as specific and verifiable measures of outcome resulting from an activity. They are normally selected on the basis of the following criteria:

(1) Objectivity – agreement about the value of the indicator should be easy so that different individuals using the same information will come to the same conclusion.
(2) Appropriateness regarding the policies of the programme as well as the local situation.
(3) Validity – the indicator must correctly reflect the concept behind it so that it can become the basis of making decisions.
(4) Measurability – the indicator should be easy to measure from available information.

When quality standards are not being met the possible causes should be sought, and corrective action evolved in discussion with the unit manager and the staff concerned. This may involve not only training and supervision but also re-organisation of services, social marketing in the case of promotive services, logistics, and so on.

Standards of performance allow for self-assessment, assessment by the peer group, and assessment by the supervisors. They help to make the process of assessment mutually objective.

SAFETY STANDARDS

With infections such as Hepatitis B and HIV becoming important sources of danger for health workers, occupational health and safety standards require particular attention. There are, of course, issues other than risk of infection such as long working hours which lead to fatigue and reduced performance through reduced efficiency. Reducing the risk to both patient and health provider should be the aim.

The safety standards commonly recommended cover the following situations:

Personal precaution   What to do when the skin on the hands, forearms and exposed body parts has cuts, is grazed, chapped or inflamed.
What to do when there is a risk of contact with blood or body fluids of patients.
How to handle sharp instruments.

Accident prevention   Actions to take when one's skin has been splashed with blood or body fluid of patient.
Actions to take when there is an accidental cut or needle-stick injury during a procedure.

Waste disposal   How to dispose of contaminated material, gloves and equipment. In particular, how to make it safe from those, especially women and children, who recycle waste.

Re-use of equipment   Methods of sterilisation, especially of 'disposables', which are used over and over again for economy.

Tables 5.3 to 5.6 give details of precautions in specific circumstances.

- Clinical work with individual patients
- Work in hospital, Health Centres and dispensaries

Table 5.3 **Precautions recommended for all staff**

Invasive procedures in all patients
- Have vaccination against hepatitis B.
- Cover all cuts and abrasions with waterproof dressing.
- Never pass sharps hand to hand. Use a tray instead.
- Do not guide needles with fingers when suturing.
- Do not resheath needles after use.
- Dispose of all sharps safely into approved containers.
- Put disposables and waste into marked waste bags for incineration.

Additional precautions when caring for known HIV and hepatitis B positive and high risk patients.
- Consider non-operative management.
- Remove all unnecessary equipment from the theatre.
- Observe highest level of theatre discipline.
- Have only experienced staff in the theatre.
- Use double glove, mask, eye protection, boots, impervious gowns, closed wound drainage.
- Disinfect theatre floor with hypochlorite.

Table 5.4 Ways to avoid exposure to HIV and hepatitis B

- Apply basic hygienic practice with regular hand washing.
- Cover existing wounds and skin lesions with waterproof dressing.
- Take simple protective measure to avoid contamination of people and clothing with blood.
- Protect mucous membrane of eyes, mouth and nose from blood splashes.
- Take care to prevent wounds, cuts, and abrasions in presence of blood.
- Avoid use of sharps whenever possible.
- Ensure safe handling and disposal of sharps.
- Clear spillage of blood promptly and disinfect surfaces.
- Ensure safe disposal of contaminated waste.

Table 5.5 Invasive procedures to be avoided by health workers who are HIV or hepatitis antigen e positive

- Surgical entry into tissues, cavities or organs.
- Repair of major traumatic injuries.
- Cardiac catheterisation and angiography.
- Deliveries, caesarean sections or other obstetric procedures during which bleeding may occur.
- Manipulation, cutting or removal of any oral or perioral tissue including tooth structure, during which bleeding may occur.

Table 5.6 Safety policy for operating theatres

Full precautions are indicated when following risk is known:
- Homosexual or bisexual males.
- Intravenous drug abusers.
- Persons with history of sexual contact with partners from an area of high HIV or hepatitis B prevalence.
- Recipients of unscreened blood transfusion in areas of high HIV or hepatitis B prevalence.
- Haemophiliacs who have received unscreened blood products.
- Known HIV or hepatitis B positive patients.
- Sexual partners of any of the above.
- Children born to seropositive mothers.

Full precautions to be taken for following surgical procedures.
- Emergency surgery for major abdominal or orthopaedic conditions and burns.
- High risk elective surgery like major abdominal, gynaecological and cardiovascular operations; orthopaedic operations requiring power tools.

PERSONAL SKILLS OF THE MANAGER

Managing time

As managers, members of the District Health Team are concerned with the effective management of the health programme of the District as a whole. But management must start with the effective personal management of each member of the team. Their own time and the use they make of it is one of the single most important assets to the District.

Generally speaking, Health Managers have learnt by experience during their professional training how to plan and use time in order to study and fulfil clinical commitments as well as maintain social and family life. Now, as managers, a new aspect of managing time becomes important. There can be a constant conflict between professional and managerial roles. By inclination and training physicians may be equipped to treat and care for individuals, yet the job requires them to have a concern for the health of communities and increasingly to manage the work of others rather than do clinical work themselves. A first step to resolving the conflict is to analyse how time is actually spent. This can be done on both a long-term and a short-term basis.

On a long-term basis, consider how you have spent your time during, say, the previous 12 months, and quantify in percentage terms, as accurately as possible, how much time has been spent on, for example:

- Clinical work with individual patients
- Work in hospital, Health Centres and dispensaries
- Time spent in planning and assessing health needs in the community
- Time spent with other health workers, helping them to do their work better
- Travelling
- Management and administrative tasks
- Study and keeping up to date professionally
- Personal and free time

Then think carefully about the nature of your job priorities in the Health District, the most effective contribution you can make, and priorities for your own work. This should result in a changed set of percentage time for most of the elements into which you have broken down your work. Many doctors who have done this exercise have found that they need to give more attention to the development and managerial aspects of their work than to clinical and routine administrative work. The problem is how to do it and where to find the time!

There are a number of ways of coping with the problem:

(1) Learn to say 'no'. Many of the demands on time are made by other

people, and ways can be found to say 'no' or to minimise those demands if they do not contribute to the achievement of your own priorities. Unfortunately it is often easier to respond to other people's demands than to establish one's own priorities and work on those. Of course, some demands are important and people must not be sent away without due consideration. Here a simple motto to follow is: 'Be gracious with people, but ruthless with time'.

(2) Delegate. Effective managers ask of every piece of work that comes to them 'who else could do this?'. They do not work themselves out of a job in this way, but create time to concentrate on the more important aspects of their work, and also create opportunities for others to take on more responsibility.

(3) Make arrangements with colleagues for rational distribution of work. If the problem is that too much time is being spent in doing clinical work in the main hospital in the District, then it may mean starting discussions with other doctors in the hospital or with the hospital authorities about the importance of rural health programmes in relation to hospital work, and agreeing new patterns of work.

(4) Plan specific 'periods' of time to do specific jobs – for example, for the first half-hour of every Monday plan the week's work with colleagues, and for the first three weeks in the month spend every Friday afternoon developing the training strategy for the District.

(5) Develop a heightened awareness of time through use of a systematic 'Diary Review', such as that shown in Figure 5.1. Note down systematically over a two-week period, every half-hour during the working day, precisely what you are doing, and then analyse the results. You will probably be surprised by how you have actually spent your time compared with how you think you have spent it. A change in daily and weekly allocations of time should follow.

(6) Have monthly, weekly and daily lists of priorities. Many managers write down each morning a 'Things to do' list and keep it on their desk or in their pocket marking items as A, B or C priorities. By working on priority A items as far as possible despite the inevitable interruptions, they usually end each day having achieved most, if not all, of the important things they had hoped to do.

(7) Have the question 'How could I *now* be best using time?' frequently in mind. This can prevent time being spent on unimportant matters at the expense of more productive concerns.

(8) Help other workers in the District to examine critically the use of their time, so that they pay maximum attention to the most important aspects of the health programme. Time studies at clinics may identify much more effective ways for clinics to be organised and health workers to spend their time. Time studies in themselves can also be a means of evaluating Health Centre work.

Practical management 185

USING TIME WELL

'To do' list	Review of time
TODAY I MUST: 1. 2. 3. 4. 5. 6. 7. 8. 9. 10.	ELEMENTS OF MY JOB — CLINICAL, PLANNING, SUPERVISION, TRAVEL • % of my time spent in last 12 months • % of time I plan to spend in next 12 months

CONSTRAINTS	DEMANDS	CHOICES
Things I am unable to do	Things I must do	Things I can choose to do
• • • • •	• • • • •	• • • • •

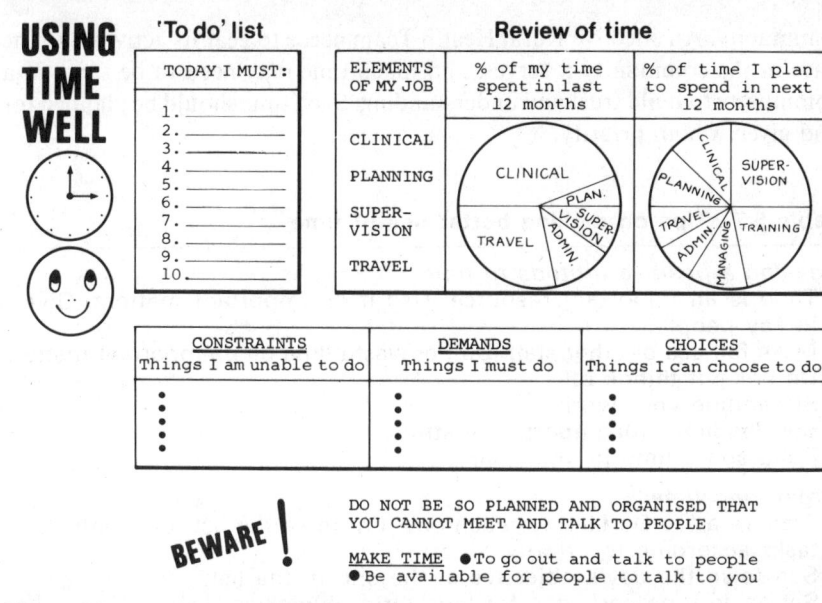

BEWARE!

DO NOT BE SO PLANNED AND ORGANISED THAT YOU CANNOT MEET AND TALK TO PEOPLE

MAKE TIME ● To go out and talk to people
● Be available for people to talk to you

Figure 5.1   Using time well

(10) Examine travelling time. Unnecessary travel not only wastes time but also keeps vehicles on the road more often than is necessary resulting in excessive use of fuel and more wear and tear. If travel is co-ordinated, two or more people can travel in one vehicle if they have assignments to perform at the same place or on the same route. Moreover, members of staff of the same health facility can have consultations and discuss problems among themselves during group travel.

(11) Think carefully about meetings. The larger the meeting, the more time is being spent at it, therefore make sure that all meetings serve a useful purpose. Have an agenda circulated well ahead in time and for important matters ask for discussion papers to be prepared. Develop also the skills needed for effective meetings so that time is not wasted on minor issues and serious matter is not neglected for lack of time.

(12) Analyse your time into Demands (things I must do), Constraints (things I am not able to do) and Choices (things I can choose to do). Consider what a wide range of choices may be open to you, before finally deciding how to spend your time.

A final word of caution, however, is needed. A rural community can have its own pace of life which is very different from that of a brisk and bustling urban community. An effective Rural Health Team needs to gear its activities to the pace of the community it serves, and much time may need to be spent in a community to build trust and understanding. Such time should be planned for and given a high priority.

Table 5.7 **Tips on making better use of time**

Develop a positive attitude to time.
- Time is an important resource. Use it on important matters. Give it to key people.
- Make full use of other staff. Do not waste time on unimportant matters.
- Do not put things off.
- Streamline your work.
- Set deadlines for important matters.
- Build spare time in your diary.

Organising time.
- Prepare a 'To Do List' for each day. Prioritise the list. Concentrate on tasks according to priority.
- Set down the day's objective at the end of the list.
- Set aside time each day for long-term objectives.
- At the end of the day, review the list and think of next day's work.
- Use a diary. Run your day by appointments and stick to them. Take diary to meetings and set down dates. Review the diary at the end of the day to see how important matters were dealt with.

Run meetings effectively. Badly run meetings are time wasters.
- Have an agenda. Prepare for each item beforehand.
- Keep meetings short with strict starting and finishing times.
- Give people a chance to contribute, but keep discussions to the point.
- Avoid 'large' meetings. They tend to go off the track easily.
- Be clear about the purpose: inform; instruct; plan; decide; negotiate.

Bring efficiency into teamwork.
- Set time for team activities e.g. field visits; review meetings; training sessions.
- Adapt membership of the team in line with the task in hand.
- Have informal contacts e.g. coffee-breaks; corridors.

Have a strategy for dealing with interruptions.
- A secretary to deal with staff matters; telephones in the midst of meetings; minor matters.
- Leave time each day for dealing with emergencies or unplanned visitors.
- Have dedicated times for casual callers.
- Turn interruptions into quickies.
- Have a 'quiet haven' to work uninterrupted if there is something important to do.

Beware of time wasters.
- Telephones / Interruptions / Poorly planned activities / Putting things off until tomorrow / Last minute panic / Poor concentration / Socialising / Trying to take on too much / Inability to say 'No'.

Delegation

Reference has been made to this important aspect of a manager's work in chapter 3. It is proposed to discuss delegation in greater detail now.

A District Medical Officer or manager of the health team responsible for all the work of a District must of necessity delegate a great deal of work to others in the organisation. The skills of effective delegation are high-order skills for senior managers, and need to be practised continually if the Health District as an organisation is to run smoothly.

The advantages of delegation are that it gives to those who delegate more time to do the things which only they can do. It makes fuller use of people's skills and abilities which would otherwise be wasted. It enables subordinates to identify themselves more closely with the aims of the organisation, and thus to feel more responsible. It increases the total capacity of the organisation without increasing resources. It gives staff generally more job satisfaction and opportunity to learn, and it ensures that there are capable people able to act in the manager's absence (see figure 5.2).

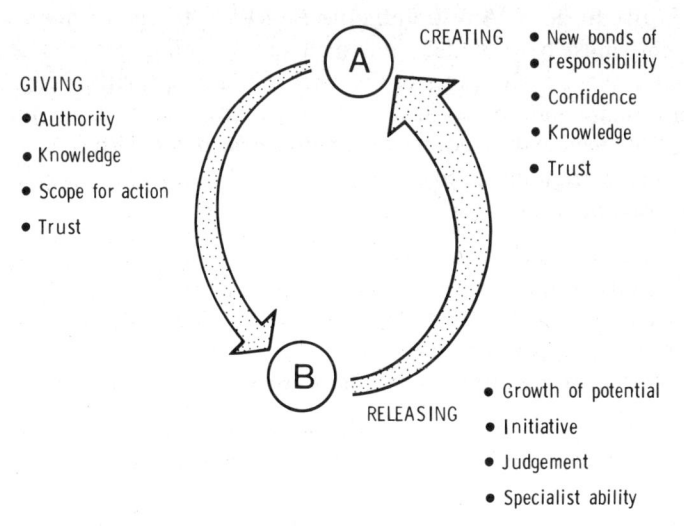

GIVING
- Authority
- Knowledge
- Scope for action
- Trust

CREATING
- New bonds of responsibility
- Confidence
- Knowledge
- Trust

RELEASING
- Growth of potential
- Initiative
- Judgement
- Specialist ability

Figure 5.2   What happens when A delegates successfully to B

There are risks in delegation as in everything else, and these deter senior staff from delegating more work and authority to their subordinates. A senior manager is still accountable for the proper performance of those tasks which have been delegated to a subordinate. For example, a District Medical Officer may in practice delegate the day-to-day running of a Health Centre or a dispensary to a Primary Health Worker who is on the spot, but is still accountable to his superiors for the services provided in the Centre. A golden rule is that a manager can delegate work and authority to carry out work, but he can never delegate his own responsibility and accountability for that work – the buck stops with the senior manager. The manager then is still responsible for mistakes made by a subordinate, yet may feel loss of control over the work, at the same time being ignorant of much of the details of the work of which he was previously knowledgeable. There may also be a deeper fear that the subordinate will become more competent at certain aspects of the work – there is evidence for example that medical assistants can be more efficient at diagnosing certain common conditions by the use of algorithms or flow charts than doctors using more established and conventional methods of diagnosis. Again, the senior manager may feel under pressure from other staff to delegate to them more responsibility before they are ready for it.

Certain conditions need to exist for effective delegation to occur. There needs to be a relationship of trust between the persons concerned. The person delegating must feel sure that the person to whom work is being delegated is competent; the person being delegated to must feel free to ask for help when it is needed. The work being delegated and what is to be achieved by it must be clearly defined to the satisfaction of both parties, and the subordinate given full authority to do it. Any training necessary for the performance of new assignments must be provided, and such supervision as may be necessary, particularly in the early stages of the work, should be forthcoming. But the most important condition is for the subordinate to feel free to get on with the work without feeling over-controlled or constrained, and to receive full credit for the work when it has been done well. The subordinate really asks for five things of his superiors:

(1) Agree with me clearly what I should do.
(2) Give me a real chance to do it.
(3) Give me knowledge of my progress.
(4) Give me help when I need it.
(5) Give me recognition when I have done it.

Controls and methods of reporting back should be agreed. This can be done by agreeing what standards of work apply, and agreeing how the work is to be checked against those standards. Then when variations in standard occur, the causes of the variation can be identified and corrective action taken to remedy the situation. An example would be where a health worker in charge of a rural dispensary is delegated authority to dispense a certain drug up to a limit of, say, 1000 mg a week. Were he to prescribe more than that amount in any one week, the variation would be sanctioned by a more senior health worker, and the reasons investigated. It could be that the health worker is dispensing too much for each patient, or for the wrong condition, or that there is a genuine increase in need for that treatment. Whatever the reason, action can be taken either to give more instruction or training to the health worker or, if necessary, to alter the standard of 1000 mg. Such a system of delegation and control can

Table 5.8 **What is your style of delegation?**

1   A DO-IT-ALL type
    This person does no delegation. When overburdened with work the trivial functions get passed on to others. Such a manager believes that only he knows enough or is competent enough to make decisions. As a result he gets overloaded.
2   A KNOW-IT-ALL type
    Work is passed on to subordinates but the manager constantly interferes and interrupts them in their work.
3   A DELEGATE-IT-ALL type
    Work is passed on indiscriminately. The manager hopes that the subordinates will accomplish it to the best of their abilities.
4   A THOUGHTFUL type
    The manager considers what needs to be done, who should do it, gives them the means to do it and maintains contact with the progress achieved.

be a powerful deterrent to malpractices such as misuse or selling of drugs. It gives the health workers freedom to work within limits which are reasonable so far as their training and experience are concerned.

Ground rules to follow in delegation

(1) Do not delegate work because it is tedious and distasteful. Delegation is meant to create motivation and for the development of staff.
(2) For effective delegation:
    . Define and agree the task, the method to be followed and the results expected.
    . Give training whenever necessary. Task analysis helps to identify those components of the task which are not being performed well.
    . Clarify authority and inform others about it.
    . Agree specific outcomes and the standards expected.
    . Agree the procedures for reporting back.
    . Feedback should be immediate, specific with constructive suggestions, and encouraging (see table 5.8).

TEAMWORK

In chapter 4 a District organisation was looked at as a skill pyramid, as a series of interrelated groups, and as a network of relationships. The success of the health activities in the District depends to a large extent on people working well together in small groups, and those small groups relating well to one another.

In an established District there can be anywhere between 40 and 100 working teams (see table 5.9) and senior management will have a direct

Table 5.9 **Pattern of organisation in an established district**

Description	Number	Approximate no. in working group	Membership of working group	Function/purpose
Sub-centres	20–80	3–6	Rural dispensary assistant Village health worker (male) Village health worker (female) (Traditional birth attendants (2)) Cleaner Nurse aide	First level of Health Care: : Community development : Diagnosis : Treatment : Referral to Health Centre : Health education : Immunisations : Sanitation
Health Centres	2–4	8–11	Medical assistant/Doctor Rural medical aid Village midwife 2 Community nurses Lab. technician Sanitarian Registration clerk 3 Cleaners/Aux. Staff	Second level of Health Care
District Mobile Team	1	5–6	District Medical Officer Nurse/Midwife 2 Medical recorders 1 Nurse student Driver	Supervision of Health Centres Clinics Health education

District Management Team	1	3–4	District Medical Officer Matron Hospital Administrator Hospital Finance Officer	Management of all District Health Services Executive of Health Authority
Hospital Management Board	1	12+		
Hospital Doctors	1	3–4	District Medical Officer 1–2 Physicians 1–2 Surgeons	Diagnosis & treatment Training & support to all health workers
Hospital Wards & Departments	10	6–8	Ward sister Physician or Surgeon Nurses Auxiliaries Attendants/Cleaners	Care & treatment of patients

involvement or interest in nearly all of them. Where the organisational structure is hierarchical it becomes critically important to try and organise horizontal relationships between the different agencies, through a process of team work. Orientation is needed for various staff to help them co-ordinate their efforts to bring about an improvement in the health status of the community as a whole. Definition of tasks is helpful for plotting out exactly where activities need to be co-ordinated. Training sessions are needed to help existing health providers acquire the skills necessary for performing specified tasks, the aim being to produce practical multi-purpose health providers rather than uni-purpose ones, trained as much together as possible and within the environment in which they are going to work. Managers who take on this task of team-building will recognise the following features of effective working teams:

(1) A clear *purpose* and *common task* which everyone in the group understands and is committed to. Managers need to help each group to appreciate what its own function is and how it fits into the overall work of the District.
(2) Each member of the group has a *clear idea of his or her own job* and how it relates to the work of others.
(3) Individual members of the group *understand the work and duties of others*, particularly where there is overlap in functions (for example, a nurse and medical assistant may each do similar work from time to time). Where members must be able to replace one another or overlap in their functions, this understanding is strengthened by having multi-purpose workers, particularly at village and Health Centre levels.
(4) *Flexibility* between members is helpful so that the work of the team does not collapse when one member is absent.
(5) This suggests that a good deal of *learning* and *training will go on* within the team, encouraged and stimulated by the team leader.
(6) *Leadership*. In most working teams the leader is clearly identifiable as the person in charge. In some teams (for example, a working group set up to achieve a specific task) the leader may not be formally appointed, but someone will need to take on the leadership function, even if the leadership role changes depending on the task in hand. For instance, a doctor may normally be the leader of an outreach team but when the same team goes out to run antenatal clinics it may be a midwife or community nurse who takes charge. Good leaders are recognised as such by the rest of the team.
(7) *Stability and continuity*. If the members of a group continue changing there can hardly ever be teamwork. On the other hand, a group which never changes its membership may become set in its ways and complacent.
(8) An effective group needs sufficient *resources* to carry out its task (these

need not be costly as is the case with much of Primary Health Care) and it needs its own *working methods* and *procedures* which are well understood and practised.

(9) Good *relationships* within the group are vital and require openness, understanding and a willingness to help. Amongst the team manager's most important skills are those of human relations – the art of building and strengthening good relationships with and between others.

(10) The real test of a team's success is its results, so it is important to have ways of measuring success and recognising achievement.

(11) *Loyalty*. An effective team develops a strong sense of cohesion and loyalty which enables it to work well and often to tackle new problems successfully. But this loyalty should not be at the expense of other groups. Groups need to co-operate with others, and frequently people will be members of more than one group (for example, a nurse who works in both a Health Centre and with other nurses in the District). The District Medical Officer has a key part to play in overcoming the dangers of groups working in isolation by explaining and interpreting the functions of particular parts of the health organisation to other parts, and seeing that the work of the total organisation is well balanced and understood.

This can be done through 'linking-pins' in the organisation. Groups working at different levels in a District can be linked together through the involvement of one member in each group in the work of another group at a different level in the organisation. Thus, each Rural Dispensary Aide (RDA) is a leader of health workers at the village level, and may also be made a member of a team at the Health Centre, headed by a Medical Assistant, which includes other Rural Dispensary Aides and key Health Centre staff. The Rural Dispensary Aides thus provide a vital link between the two levels of health care – to ensure that Health Centre staff understand the needs of those working at the village level, and that members of the village teams understand how the Health Centre works and deals with patients referred from the village. The Rural Dispensary Aides should also from time to time meet with 'colleagues' working in other villages to share ideas on how they run their dispensaries. Similarly, Rural Medical Assistants act as a link between Health Centre and District levels of care.

What makes a good team?

In a good team a balanced mix of skills and personalities is needed as follows:

The ideas person – The source of all sorts of ideas about what should be done, and how problems could be solved. Of course not all the ideas would work.

The contact person – Gets on well with all sorts of people and knows everybody. Good at explaining the work of the team to outsiders and good at getting support.

Table 5.10 Assessment of team effectiveness

The effectiveness or otherwise of a team may be assessed by rating the following factors on a scale from 1 to 7:

Team objectives Not understood by team	1 2 3 4 5 6 7	Clearly understood
Team is negative towards objectives	1 2 3 4 5 6 7	Team is committed
Utilisation of member resources Our abilities, knowledge and experience are not utilised fully	1 2 3 4 5 6 7	Our abilities, knowledge and experience are fully utilised
Degree of mutual support High suspicion	1 2 3 4 5 6 7	High trust
Control methods Control is imposed on us	1 2 3 4 5 6 7	We control ourselves
Handling team conflict We deny, avoid or suppress conflict	1 2 3 4 5 6 7	We accept conflicts and work them through
Experiential learning We ignore and do not learn from team experience	1 2 3 4 5 6 7	We analyse our experience and learn from it
Team environment Restrictive; pressure to conform	1 2 3 4 5 6 7	Free, supportive, respect for differences
Communications Guarded, cautious	1 2 3 4 5 6 7	Open, authentic
We do not listen to each other	1 2 3 4 5 6 7	We listen, understand and are understood
Sense of belonging We have no sense of belonging	1 2 3 4 5 6 7	We feel we belong

The smoother – Understands everyone else's point of view and has genuine sympathy with people on both sides of the argument. Good at maintaining relationships between other members of the team.

The finisher – Gets things done. He keeps a deadline and makes sure that the details are attended to. He is also particular about knowing what decisions are reached after a discussion.

The rules person – Knows all the regulations and can use this knowledge either to find solutions to the problems, or to block other people's initiative.

The chairman – The one who can identify the main points of an argument, find out areas where the team members are in agreement, and find acceptable compromises.

Most people are a mix of these traits in varying amounts with one type of personality predominating.

Table 5.11 **Techniques for efficient teamwork**

1. Set time for team activities – field visits, training sessions, review meetings.
2. Share information on what you are all doing.
3. Consider how to make time outside teamwork for working as groups (of two or three), or individually, and with other groups and individuals outside the team.
4. Informal contacts e.g. coffee time, corridors.
5. Adapt membership of the team in line with the task in hand.
6. Management walkabouts.

Group work

Often it is necessary to form a small working party or group within a team or between teams, to address a particular problem. How large should such a group be? Obviously, the larger the group the more is the talent represented. On the other hand small groups develop a strong commitment to the work in hand. It is the general experience that for efficient functioning the ideal size is between three and nine.

Membership of the group should be by talent, knowledge of the problem or piece of work to be addressed, and above all, the ability to work with others. At least one member of the group should be primarily concerned with *task*, and one with the *process*.

The way groups perform depends on the nature of the task and the personalities of the participants.

Nature of the task

Groups which are committed to the tasks and believe they are competent to do it find it easy to work in a participative, democratic manner. Directive leadership is needed if:

- the task is not perceived as important or interesting;
- members of the group feel unskilled or incompetent for the task.

Personalities of the participants

The working styles of individuals depend not only on nature and nurture but also on the predominant culture, training and experience amongst a host of other influences. In group work individuals tend to show a mix of styles with one of the following predominating:

Type	How to handle
Positive	Great help in discussions. Use him or her frequently, and let his contributions add up.
Negative	Play on his ambitions. Recognise his knowledge and experience and use them.
Know-all	Let the group deal with his expositions.
Quarrelsome	Do not get involved. Keep your cool but stop him monopolising.
Loquacious	Interrupt tactfully. Limit his speaking time.
Shy	Address questions to him. Increase his self confidence. Give credit when possible.
Uninterested	Ask him about his work. Draw him in and get him involved.
Highbrow	Do not criticise. Use the 'yes – but' technique.
Persistent questioner	Tries hard to trap the group leader. Pass his questions back to the group.

The working styles described above are caricatures of human nature. Bearing these frailties in mind working groups require eight roles to be performed during their sessions to be productive. These roles are listed below. A group without someone taking on these functions will perform sub-optimally. If need be one person may perform a dual role, but not more than two.

Chairman	Good listener and judge. The qualities required are those of being disciplined, focused and balanced rather than brilliant.
Shaper	Achievement oriented, and full of drive. Provides the stimulus and leads the pack.
Innovator	Intelligent. Can be sensitive to criticism.
Monitor/Evaluator	Has a critical mind. Can see the flaw in an argument.

Resource person	Brings new contacts to the group.
Implementer	Turns concepts and ideas into tasks.
Team worker	Good at listening and building bridges.
Finisher	Keeps an eye on deadlines, schedules and completion dates.

However well people working in the same organisation may know one another, group dynamics take time. All groups go through the cycle of Forming – Storming – Norming, before they can begin Performing.

Forming stage	People are polite, impersonal, guarded and watchful.
Storming stage	Difficulties and confrontations arise. Conflicts need managing. Members may feel stuck with no progress being made. Some get demotivated. Some may opt out.
Norming stage	Getting organised, and establishing systems and work procedures.
Performing stage	Group maturity and resourcefulness develops. Members are open, supportive, tolerant and sharing. The group is now beginning to produce results.

In the long run groups tend to get cosy and their effectiveness goes down. It may become necessary to dissolve and re-form groups, or from time to time bring in new members.

Meetings

Regular meetings are amongst the important tools of management. Much of the work to be done by a team is decided upon in meetings. The conduct of such meetings is essential for the success of the undertakings of the team. Handled well, they can produce great benefits. Run badly, meetings can lead to frustration, acrimony and are a waste of everybody's time. Positive personal relationships amongst those attending are important. There should exist a relationship of trust so that time is not wasted by people being unnecessarily defensive. Courtesy and good manners are vital for creating an atmosphere in which free exchange of views can take place.

Managers use meetings for:

- providing information;
- seeking opinions;
- reaching decisions;
- preparing documents; or
- a combination of the above.

Many meetings flounder because little thought has gone into them either beforehand or during the meeting itself. There are some simple tips which can help run meetings successfully. These can be summarised as the 5 P's:

- Planning
- Pre-notification
- Preparation
- Processing
- Putting it on record

Planning

Think through the purpose of the meeting in advance, and what it is intended to achieve. Is the meeting intended mainly to inform, instruct, decide, negotiate, or a mixture of any of these?

Setting clear objectives and deciding about the size and membership of the meeting are part of planning. The objectives need to be carefully thought through and reviewed repeatedly prior to the meeting. The objectives help to determine the composition of the membership. It is often said that the effectiveness of a meeting is inversely related to the number of people attending it. Nevertheless, a balance must be struck between adequate coverage and keeping the size small. If representatives of groups are selected to attend, then ample opportunity must be provided for groups to debate the issues beforehand so that their views are expressed at the meeting by those representing them.

One essential aspect of planning is to decide beforehand where you want to be at the end of the meeting. In other words, what views must be obtained, and what decisions made before the end of the meeting.

Pre-notification

Inform other members of the meeting about what is to be discussed and why, so that they can come prepared.

Preparation

Arrange an agenda in its proper sequence and allot the correct amount of time for each subject. Give more time to those issues which are more important.

The length of the agenda is critical so that discussion can take place within the allotted time. If there are too many items the Chairman must devise an alternative strategy for dealing with them.

Processing

Structure the discussion of each item to keep members to the point, and to avoid the pitfalls of repeatedly covering old ground under 'Matters Arising'.

This section is for reporting back on action taken since the previous meeting. It is for the chairman to identify key issues, make a summary and give a succinct report. The aim should be to get matters resolved so that they cease to become running items from one meeting to another meeting. Develop an awareness of group behaviour, control private discussions, and reconcile disagreements within the group. Give people a chance to contribute, but keep discussions to the point. At the end of the meeting the participants should feel that they have had ample opportunity to voice their views.

The effective use of time is the hallmark of good chairmanship. Without prior planning and adequate preparation the meeting can easily run out of time before decisions are reached. Obtaining decisions on the items in the agenda is the main purpose of the meeting, so that the next set of actions can be decided even though the matter may not have been completely resolved.

Putting it on record

Summarise and record decisions made and action to be taken. It is important that all participants should go away with the same impression about what conclusions were reached.

The record of the meeting should not only show the decisions arrived at, but also carry an action plan stating *what* is to be done, *who* has been assigned to do it, and the deadline agreed, that is *when*.

In between meetings good teamwork can be judged by the extent to which team members work together to achieve specific tasks, share information or network of contacts, and help each other to solve problems.

MOTIVATION

One of the most important resources available to the Health District is the working potential of those employed within it. But it is often said that those involved in rural health care are under-employed, disinterested in their work, and not prepared to put into their work more than the minimum that is required.

The motivation of health workers working with disadvantaged rural and urban communities is important since the attitudes of those in positions of leadership have a considerable influence on the rest of the team. Lists can be drawn up of factors about which health workers have positive attitudes (motivators), and a list of factors about which they feel badly. Such a list is shown in table 5.12 as one example. Obviously, the factors will vary from place to place. But the implications for a District Health Manager are clear. If the negative aspects of the job remain, the morale and enthusiasm of the health workers and subordinate staff will deteriorate. Action must be taken to remedy the negative aspects (for example, better arrangements for drug

Table 5.12 **Factors associated with motivation**

Factors associated with high job satisfaction	What can be done to increase job satisfaction?
Having a job with an important purpose especially when the impact is immediate.	• Identify important problems with the staff. • Agree on what is to be achieved. • Involve staff in reviewing their job for better results.
Having opportunities to use skills.	• Jobs should provide opportunities to use a variety of skills. • Staff should have the freedom to organise their own activities and time. • Encourage staff to achieve agreed objectives.
Getting recognition for good work.	• Review work done with staff. • Discuss progress, difficulties and give advice. • Complement their good work in writing, and mention it in the annual reports.
Having opportunities for advancement.	• Recommend efficient staff for promotion. • Recommend staff for further training. • Encourage staff to participate in regional or national workshops and seminars.
Good supervision.	• Make sure the supervisors are well trained. • Provide support to outlying areas. • Motivate supervisors.
Reasonable working conditions.	• Maintain clean and pleasant work places. • Refund travel and transport allowances on time.
Good working relationships.	• Help staff to settle conflicts amicably. • Help staff with personal problems when approached.
Reasonable pay.	• Ensure salaries are paid on time. • Deal with administrative incompetence.

Practical management 201

Table 5.12   *(contd.)*

Factors associated with poor job satisfaction	
Isolation.	• Arrange for regular supervisory visits. • Give preference in selection for refresher courses. • Regular newsletters.
Having to work without proper supplies of drugs, equipment etc.	• Have an Essential Drugs Policy. • Prepare guidelines for appropriate prescription of drugs.
Fear of making mistakes which could affect patients' lives.	• Visit regularly for teaching and consultation sessions.
Boredom and routine.	• Have defined targets and give praise when achieved. • Analyse shortcomings sympathetically. • Encourage regular attendance at meetings, training sessions and participation in distance learning.
Conflicting pressures (e.g. DMO to limit over prescribing, and from patients to provide a pill for every ill).	• Help establish mechanisms of social control e.g. Health committees, and other similar activities of community involvement.

supplies, refresher courses, better referral back-up, and so on). At the same time the positive aspects of the job need to be reinforced, as these are the aspects which encourage the health workers to work with enthusiasm.

What promotes motivation amongst workers in an organisation?

A great deal has been written about motivation in the work force. All organisations, whether commercial or otherwise, are live communities with a common purpose. They are made up of free citizens with value systems, rights, and minds of their own. Consensus amongst management and the work force is important if the organisation is to move forward. Consensus in the following five key areas is especially important:

(1) The mission – Hence the importance of the mission statement.
(2) The goals – This relates to a definition of specific goals for different level workers and thorough discussion about them.
(3) The means to accomplish the goals, including reward and incentive systems.

(4) The means of measuring progress including quality standards, reporting and feedback.
(5) The strategies for what to do when things go wrong.

Secondly, organisations are people. Individuals constantly rub shoulders with each other. They have to work in groups most of the time, and in the nature of things, some will have more power than others. Thus group psychology becomes an important determinant.

Thirdly, at the individual level we all bring an unstated, unconscious psychological contract into every situation, be it family, social group, or work. Unlike a formal contract the psychological contract is never written down. In it the individual offers to give something in return for something received. Motivation happens when the contract is balanced. When it is unbalanced, one side or the other feels cheated. When motivation happens there is excitement, energy, effort and enthusiasm. In other words, the release of 'E' factors. The balance of energy contributed and expectations met makes up the psychological contract. Each of us continually makes calculations balancing the 'E' we have to expend, and the results of that expenditure against satisfying something within us. In the work situation the outcome of such a calculation can be:

Co-operative – One is there because one agrees with the goals of the organisation, and the people in it.
Monetarist – One works because one is paid to work, and won't do more unless paid to do more.
Coercive – One is forced to do things (for example police, army, hospitals etc.).

It will appear that motivation is something very complex. Its roots lie in the social and behavioural driving force of the organisation, as well as the psychological make-up of the individual employee. The ways in which managers view the people they manage will largely determine the level of motivation in the work force. A variety of models have been put forward to explain the manager-workforce interaction as follows:

(1) *The rational economic model*, also called Theory X. This model states that most people have to be enticed and coerced into work by means of economic incentives, and require close supervision. The arguments based on Theory X are further developed below.

Theory X (see table 5.13)

(a) The average human being has an inherent dislike of work and will avoid it if he can.
(b) Because of this human characteristic of dislike of work, most people

must be coerced, controlled, directed, or threatened with punishment to get them to put forth adequate effort towards the achievement of organisational objectives.
(c) The average human being prefers to be directed, wishes to avoid responsibility, has relatively little ambition, and wants security above all.

(2) *Self-actualising model*. According to this model every individual has a hierarchy of needs. An understanding of individual needs is essential for developing the worker's potential. This is also called Theory Y. These concepts are further explained below.

A hierarchy of needs

Once an individual's basic needs (food, clothing, shelter, warmth, safe environment) are met, the higher needs for love, esteem and fulfilment of personal potential are released. None of these is absolute as is the case with the basic physiological needs. As soon as one is satisfied, the matter of its satisfaction ceases to be important. Also a want changes in the act of being satisfied. As a want approaches satiety, its capacity to reward and with that its power as an incentive diminishes fast. But its capacity to deter, to create dissatisfaction and act as a disincentive rapidly increases.

Theory Y (see table 5.13)
(a) Work is as natural as play or rest. Depending upon *controllable conditions*, it may be a source of satisfaction (and will be voluntarily performed) or a source of revulsion (and will be avoided if possible).
(b) External control and punishment are not the only means of achieving effort. Man will exercise self-direction and self-control in the services of objectives to which he is committed.
(c) Personal satisfaction can be the direct product of effort directed towards organisational objectives.
(d) Most people, under proper conditions, learn not only to accept but to seek responsibility. Avoidance of responsibility, lack of ambition, and emphasis *on security are generally consequences of experience and lack of it but not inherent human characteristics.*
(e) The capacity to exercise creativity in solving problems is widely distributed in the population. But under modern working conditions, the intellectual potential of the average human being is only partially utilised.

(3) *Social model*. The individual worker has a number of other needs than purely economic. The peer group and relations with co-workers has a bearing on work performance.

Table 5.13 **Two sets of attitudes to work**

Theory X
Manager

'I am not really interested in this job – My workers are lazy and idle; they only work if I threaten to discipline or "bribe" them; they do not want to better themselves; they avoid responsibility; they don't want to do anything new.'

Result: Workers sitting around, sulky, no initiative, work not done properly.
Health Centre dirty and badly organised.
Workers afraid of the manager and keep out of the way.

Theory Y
Manager

'We have a difficult job to do here but we are making some progress – My workers enjoy their work; I explain clearly what I want them to do, give them whatever help they need and praise them when they do it well; I trust them to work well on their own because I know they will do it; they get a lot of satisfaction from doing their work well – I encourage them to take more responsibility and they do so; they frequently solve their own problems and often help me solve my own; I do whatever I can to help them achieve their own goals.'

Result: Workers very active.
Well run Health Centre.
Many village activities.
Workers, managers and villagers work together on new projects.
Workers have new ideas and put them into practice.

(4) *Complex model.* An individual's motivation will be governed by many variables at different times and in different situations. Incentives change in accordance with the individual's changing perceptions.

Taking all these theories into account, what are the practical steps that one can follow? For everyday purposes it is usually enough to remember the following points:

(1) *People like targets.* Without something to aim for, work is just routine. But the targets should be the ones people have themselves determined, and look upon them as their own. Targets should be relatively short term, possibly less than six months; otherwise they are too far off to be real.
(2) *People like to feel good*, in order to work better. One feels good when one succeeds, that is one meets the target and receives praise from those one looks up to. It is easier to raise a person's work standard by raising targets and praising their achievement than by reproaching for shortcomings.

'Theory X' attitudes are quite prevalent, and particularly relate to those who have to carry out more menial tasks. How can the conditions of 'Theory Y' apply to, say, a sweeper who is employed to keep the floors and surrounding of a Health Centre or dispensary clean, hygienic and tidy? Yet there are many examples of people who have been encouraged to do this kind of work well, through praise, being helped to feel part of the Health Centre team, and having a clear idea of the standard of work required. Such people have responded by extending the limits of their work, taking on new tasks and finding new ways to be helpful. Similarly well-motivated and effective working teams often look for ways of improving their own work and look for new and challenging tasks to achieve.

District Health Managers therefore need to think carefully about their own motivation. They must have a genuine desire to improve the health of the District population; they must believe that the job is worth doing, and can be achieved; have faith in their own abilities to achieve it; and have clear goals and strategies to bring it about. They must also be able to motivate other people to share in the task. This may be done in different ways with different people – with professional colleagues it may be by working with them, persuading and 'selling' ideas in meetings, seminars and committees; with other health workers by good organisation, supporting and encouraging them; with village workers and leaders in the community by understanding, listening, co-operating and helping; and with the mass of ordinary people in the community by helping, educating and encouraging them to cope with their own lives and health problems.

Medical education is so much disease oriented that many professionals find themselves unconsciously holding the values, concepts, methods and behaviours dominant in different specialties. The focus is more on disease processes and outcomes, and less on people. Table 5.14 shows how motivation can be rated on a scale of preferences for what is of core importance and what is marginal.

The principles of accessibility, availability and acceptability on which PHC is founded require a new style of professional attitudes. These are described in table 5.15. Finally, the reader can assess his or her own motivation by reference to table 5.16.

Table 5.14 **Preferences**

Core or first	*Marginal or last*
Power	Weakness
Comfort	Discomfort
Wealth	Poverty
Core location	Peripheral location
Urban	Rural
Industrial	Agricultural
Things	People
Clean, odourless	Dirty, smelly
Uniform	Diverse
Tidy	Untidy
Controlled	Uncontrolled
Certainty	Doubt

Preferences for technology

Core or first	*Peripheral or last*
Large-scale	Small-scale
Capital-intensive	Labour-intensive
Modern	Traditional
Hardware	Software
Inorganic	Organic
Market-oriented	Subsistence-oriented
Mechanical	Human or animal-powered
Developed in core	Developed in periphery
'High' technology	'Low' technology

Table 5.15 **Old and new style professionals: preferred contacts, perceptions and roles**

	Old style	New style
Contacts preferred with people who are:	Upper class Powerful Educated Male Adult Light-skinned	Working class Weak Illiterate Female Child Dark-skinned
Clients considered to be:	Conservative/Obstinate Passive Ignorant At fault Beneficiaries Inferiors Dependent adopters	Rational Active Knowledgeable Blameless Collaborators Colleagues Independent innovators
Roles of professionals:	Teacher Expert	Learner Consultant

Table 5.16 **Recognise your own motivation to work**

Pattern A
I am likely to do my best work in situations that:
- will produce practical results, useful products.
- involve other people or take group effort.
- let me work towards goals step by step in an orderly way.
- are real and not just dealing with theory.
- give me a clear picture of what other people are doing and what they regard as important.
- have realistic schedules and do not expect too much too soon.
- let me learn from first hand experience on the job.
- let me use the practical skills and facts I possess.
- give me a regular work schedule, but give me some variety and time to socialise.
- let me work with concrete things, hands-on materials.
- let me think out loud with other people.

Pattern B
I am likely to do my best work in situations that:
- make practical sense to me.
- have a clear organisation in them.
- are practical and realistic.
- let me know just what is expected of me.
- let me work at a steady speed, step-by-step.
- require accuracy and careful attention to detail.
- require patience.

Table 5.16 *(contd.)*

- do not have many surprises in them.
- let me use my practical experience.
- let me use my memory for facts and faces.
- let me think through a problem before I have to act on it.

Pattern C

I am likely to do my best work in situations that:
- put me on my own initiative.
- let me plan and carry out new projects.
- involve other people in solving problems, such as group work.
- let me be innovative.
- let me try out my ideas to see if they work, to see how others react to them.
- do not require a detailed accounting of how I use my time.
- provide variety and a minimum of routine.
- let me figure out how to put theory into practice.
- let me make mistakes without penalties, and let me learn from mistakes.
- challenge my imagination.

Pattern D

I am likely to do my best work in situations that:
- let me work in my head with my own ideas.
- let me devise solutions in my own way.
- give me a chance to be creative.
- let me set my own standards of quality.
- let me work hard when I feel like it, and go easy when I need to.
- do not burden with too many routines.
- have important ideas behind them.
- give me ample time to think out my ideas before I have to act.
- let me use my hunches and inspiration.
- let me follow my curiosity.
- let me work in depth on things of importance to me.

COMMUNICATION

Good communication is the corner stone of any strategy of rural health care. Village health workers identify major health needs by observing, listening and talking with villagers; the relationship between patient and health worker is one of dialogue, with both parties listening and trying to understand what the other is saying; qualified staff in supervising others do not just tell them what to do, but adopt a problem-solving approach, discussing problems together and jointly agreeing solutions.

Much of medical and health worker training has been based on the assumption that those who have health knowledge are 'experts' and must pass on their 'expertise' to others. This of course is only half the story. The village health workers are the experts on their own village, knowing the history and village traditions, the personalities of the villagers and relationships between them; the particular problems such as water shortage, crop disease, and infections that villagers have to face. Workers in the Health Centre are the experts on the Health Centre and its problems, the needs of the population it serves. Health Managers need to take in this information as well as to give out their own expertise. So they need to be experts at listening as well as at passing on their expertise. This is communication. It often means working with whatever channels of communication are available.

Problems with communication

In many developing countries systems of administration including communication patterns were established during the colonial period. The objective was to maintain domination over the social and economic life of the people in favour of colonial interests. With self-rule, and especially now with changing over to Primary Health Care there is no need to perpetuate old patterns. Change in the established secretive patterns to a more open type of communication is taking place in all countries.

Any transfer of information is likely to come against one or more of the following obstacles:

Distortion by the sender – Holding back parts of the information by being economical with the truth is part of the power game bureaucrats like to play.

Distortion by the receiver – It is a human tendency to filter out information according to one's perception of the sender, or dismiss unwelcome news by casting doubts on its authenticity.

Distance – Either physical or in terms of layers of bureaucratic hierarchy, distance tends to make the signal faint.

Lack of clarity – Information transmitted in bureaucratic jargon or ambiguous terminology is difficult to follow.

The rules of good communication

- Use more than one medium. Even in a literate culture people obtain 60 per cent of their information from reading, and 40 per cent from talking.
- Keeping 'big' news secret is a bad policy. Allow some leaks, since they help to prepare the ear of the recipient.

- Encourage two-way communications. Passive listeners are usually poor listeners. Encourage questions, discussions, arguments and every form of participation. Such an approach may take a long time, but is effective.
- Avoid too many links in the information chain. There is distortion and filtering each time communication flow encounters a junction or another layer.

Direction of communication

Communication can be in one of three directions:

(1) *Downwards to subordinates* concerning:
   (a) directives and instructions related to tasks and jobs.
   (b) understanding and descriptions of tasks.
   (c) organisational procedures and practices.
   (d) feedback about performance.
   (e) policies for example values; goals; missions.
   Most commonly, downward communication is about (a) and (c), though (b), (d) and (e) need greater attention.
(2) *Upwards to superiors*. The information from workers to managers is usually asked rather than voluntarily supplied. This information is often intended for monitoring of the work and control of the workforce.
(3) *Lateral*. In this instance the flow of information is between people who are at the same level. This kind of information is intended for co-ordinating the output, and for the support of individual departments.

The psychology of communication

When two individuals communicate a great deal more happens than just the transfer of information. Their personalities and psychological make-ups also come into play. This is the Theory of Transactional Analysis (TA). Here transaction is defined as interaction between two individuals in which the pattern of thoughts, feelings and behaviour in one excites a corresponding response from the other. Each individual's pattern is determined by the traits and qualities which became imprinted as a result of childhood experiences to which have been added the rational approach to facts as well as qualities learnt from parent figures and role models. TA has been used as a basis for analysing problems in communication. Understanding the working styles of colleagues helps to design an appropriate strategy of communication (table 5.17).

Table 5.17 **Types of work styles**

Hurry-ups	They like to get a lot done quickly and are likely to make mistakes, or the quality of work may suffer. They often appear impatient. Constantly in a rush, they are unable to get to know people well.
Be perfects	The opposite of the hurry-ups, they are looking for perfection. Their work is accurate and reliable. Well organised, their projects run smoothly with careful monitoring of progress. They are poor delegators and are constantly checking everything for themselves. Because of all these qualities, they do not produce things on time.
Please people	These are the 'nice', 'understanding' types, who aim to please others. They are good team members, but lack assertion. Because they do not like to say 'No', they tend to get overloaded.
Be strongs	Emotionally detached and always calm under pressure, they are reliable, steady workers who can handle staff firmly and well. They hate admitting weaknesses, and do not normally ask for help. They are critical of self and others.
Try hards	They are enthusiastic about anything new, and so volunteer for new tasks. More committed to trying than finishing, their initial interest wears off before the end of the task.

Most people are a bit of all types, with one trait dominating. The first principle to remember is that it is easier to change how one reacts to different types than changing people. The strategies for dealing with different types are as follows:

For the hurry-ups	Help them to plan their work in stages by setting interim targets. Teach them to concentrate on some aspect of the task which they may have overlooked, before submitting their work.
For the be perfects	Tell them and others that their mistakes are not serious. Teach them to ask themselves what the real consequences of making mistakes are.
For those who want to please	Tell them that it is better to ask others what they want instead of guessing. Also at times suggest that they ask others for what they want and practice telling colleagues firmly when they are wrong
For the be strongs	Show them that it is acceptable to delegate, to allow others to guide them. They should 'be open and express their wants'.
For the try hards	Suggest that they stop volunteering. When they make a plan, tell them to make sure that finishing the task is included. Help them to do things in small doses and to give themselves a few minutes to decide whether they will do something or not.

It is useful to map out the Communication System of a District along the lines shown in table 5.18 and see how it links with information and other systems. By thinking carefully about formal communication flow, gaps and duplications may be eliminated. At the same time informal communications must never be forgotten. However much formal communication systems are improved, managers need to be aware of the vast amount of information which is communicated informally and should remain close enough to people to 'pick up' what is being said and felt around the District.

Managers find that up to 90 per cent of their working time can be spent in communicating. The essential elements of communication are listening, talking, reading and writing. Yet for most of the time they are hardly aware that they are communicating. Much of their work is accomplished by talking with individuals. Increasingly, managers need to discuss with groups of people, which entails careful listening if all the things that people may be trying to say are to be understood, and of course careful speaking if the manager is to get his own point of view over successfully. The larger the group, the more 'public speaking' skills are needed, besides those of encouraging others, reconciling different points of view and explaining complex issues. Writing skills are important, particularly in dealing with higher levels of government and other organisations. Where literacy levels are low, care must be given to simple, clear and unambiguous writing. A great deal of communication, however, takes place at a level beyond the spoken or written word, and is concerned more with feelings and emotions. It is expressed in looks, bodily gestures, silences. Skilled listeners take these factors into account at the same time as hearing the actual words which are used.

MANAGEMENT OF CHANGE

We have already discussed changes in the health organisation of a District for the implementation of a health plan on page 163. From time to time situations arise in the District which call for a change in the work pattern in a village, or a small division in the District or in one of the units of the health systems. As new health needs of a community are identified health workers have to change their approaches and strategies to meet the challenge. From time to time changes may be needed in the balance between hospital and Primary Health Care. Fresh ideas and new knowledge may require changes in the way care is being provided. Staff movements often make change essential. The District Health Manager frequently has the job of bringing about and implementing change. A balance has to be struck between trying to change too much or too little, between changing too quickly or slowly, and bringing about the desired change with the support of the community

Table 5.18 **Example of part of a communications chart for a district**

Communicator	Method	How often	Items for discussion	Purpose	Length
District Medical Officer or District Management Team	Meeting and Action Sheet	Weekly/Fortnightly	District Strategy, District Resources, Activities/Evaluation Common Problems Priorities	To ensure a concerted approach to management of the District	1-2 hrs
District Medical Officer and Health Centre Supervisor	Meeting individually and with Health Centre staff	Monthly	Previous month's work Unusual referrals to District Hospital Problems Priorities for next month	Supervision and support of Health Centre Supervisor. Link with Health Centre	2-3 hrs
Health Centre Supervisor and staff	Meeting	Weekly	Priorities Problems Allocation of work Matters to discuss with District Medical Officer	To ensure a team approach in the Health Centre	20/30 min
Health Centre Supervisor and village health workers	Meeting or individually in village	Monthly	Problems in the village Training and supervision referrals	To support village health workers and ensure Health Centre meets the needs of villagers	2-3 hrs
Health Centre Supervisor, village health worker and village development committee (& District Medical Officer for major problems or issues)	Meeting	3-monthly	Previous 3 month's work Problems Links between Health and other projects	To ensure continued support and commitment of Village Development Committee	1-2 hrs

on the one hand and consent of the workers involved on the other. Leadership must suggest and demonstrate new ways of doing things, but a good leader does not get so far ahead of people that changes and new approaches are misunderstood and resented.

People usually react to change from the standpoint of 'What's in it for me?' and it is therefore important to see changes from the point of view of those who will be affected.

Where people are resistant to change, they may show it through being less interested in work, sometimes 'opting out' of their responsibilities, sometimes refusing to accept that the change will take place, or complaining more about their work. In some cases people become much more dependent or antagonistic to supervisors or those they think are trying to bring about the changes. They may continue to do things in the same way as before, particularly if it is a long-established practice or if they have been well trained to work that way. In addition, if a number of individuals feel threatened by change, they are likely to strengthen themselves as a group to resist the change.

Many of these reactions will be found familiar. Take the case of a Health Centre which has been established for a number of years but which has failed to adapt during that period to changes in population, income and expectation of services amongst the population it serves. The result is that many people no longer come to the Health Centre – expectant mothers and 'at-risk' groups do not attend preventive and education programmes; those with minor ailments bypass the Health Centre and go to a more distant rural hospital for treatment; Health Centre staff dispense larger and larger amounts of drugs for cases which do not really need them, mainly to show that they are busy. Along comes a bright new medical officer who decides that changes are needed. Imagine the feelings and reactions of the Health Centre Supervisor whose reactions may be defensive: 'Why change? – we have always done it this way'; he may withdraw support by refusing to meet or communicate openly with the new medical officer; he may even be critical or apathetic to whatever new ideas are put forward. One can also see how such resistance would spread to other workers at the centre and to members of the community as well.

How might the medical officer respond to such resistance? An immediate response, but one not likely to succeed, might be:

- Defence – reacting to the resistance as a personal attack and asserting his authority as the medical officer,
- Persuasion – attempting simply to argue with the supervisor without recognising what feelings are involved.
- Criticising the way the Health Centre has been run in the past.
- Control – seeking to force change on the Health Centre.
- Punishment – threatening punishment or withholding rewards if the changes are not made.

A more successful approach would be to:

- *Attempt to understand* the feelings of the superintendent and other workers and avoid inappropriate actions.
- *Create an atmosphere* of confidence and trust that will minimise resistance, increase understanding and secure co-operation.
- *Communicate.* Encourage the staff to talk about their problems and discuss what actions may help to solve them. Explain why change is needed.
- *Gain the confidence* of those affected by the change through involvement and participation. The more the staff are involved in planning change, the less they will resist it.
- *Get the right timing.* Slow changes can be easier than rapid ones, since this gives people time to adjust, but it can also be unsettling if the change is too long and drawn out.
- *Reassure staff* that training and support will be available to enable them to cope with any new situations which may arise.
- *Be flexible.* Be prepared to accept modifications if these are seen to be desirable. There are usually a number of ways to achieve the same ends.
- *Succeed!* All successful handling of change engenders confidence and trust. The next change to be made will be that much easier.

INTRODUCING CHANGE

There is no one systematic approach for introducing change. Organisations and managers are different and there are many different factors to take into account. This checklist may, however, serve as a useful guide to managing change.

Recognising the need for change

(1) What is the objective of the change?
(2) What data do I have to support the need for change?
(3) Do others need to be convinced of the need for change? How can I achieve this?
(4) Am I sure that the need for change will not be met by other developments?

Planning the change

(1) Is the change programme clear, with well defined objectives?
(2) What is the best method for accomplishing the change?

Has it been tried anywhere before?
Can other methods be used, provided that the objectives are achieved?
Should a pilot change programme be tried first?
(3) Is the present organisation sufficiently flexible to accept change?
Is there shared decision-making?
(4) Is there a deadline for achieving the change?
Is this deadline flexible?
(5) Do I have the involvement and commitment of people who matter, for example, community leaders, senior officials? What other external influences need to be taken into account, for example, changes in government policy, regional health office, consultant colleague, patient demand?
(6) Can I introduce the change myself or should some other individual or group be used as the change agent?
(7) What will be the effects of the change on other parts of the organisation?
(8) Have I considered the reaction of individuals which may act as a barrier to the change?
Who are likely to support?
Perhaps I should meet them all individually to share ideas and to discover how they will respond to the proposed change.
(9) How can I create a climate of collaboration (table 5.19)?

Getting agreement

Having done the groundwork, the next step is to arrange a meeting with the entire group or health team that would be affected by the change. The aim of the meeting is to reach public agreement about proceeding with the change. Hearing everyone's views, both positive and negative, allows broader understanding of the objectives of the change and its implications. Achieving a public consensus helps to build the head of pressure needed for moving forward, and creates team cohesiveness. It is important, however, that all stakeholders should attend. The presence of a facilitator brought in from outside the organisation often helps.

The conduct of the meeting is important. Equal time should be given to each person to respond to three questions among other things:

(1) What appeals to you about this proposal?
(2) What are your concerns about it?
(3) Do you need any more information in order to reach a decision?

It helps to display the answers on a flip chart, if possible. At the end of the meeting progress is reviewed by confirming agreements reached, at the

Table 5.19 **Techniques for encouraging collaboration in innovations**

1. Be explicit about why you think it is important to involve the team member concerned.
2. Behave with trust, expecting your colleagues to be able to look at the ideas objectively. The best policy is 'Give trust to receive trust'.
3. Be equally receptive to positive and negative responses to your proposals.
   Be seen to value each person's opinion even if it contradicts your own, because agreement achieved after negative reaction has been explored is more powerful than agreement achieved after evading potential conflict.
4. When potential benefits are identified it is important that they should be seen as realistic and achievable.
5. When colleagues have identified the potential costs to themselves these must be addressed, however trivial they seem. Strategies for minimising the costs can then be explored. Colleagues often need the opportunity to express and deal with fears they have about the proposals for change.

same time identifying areas of disagreement and deciding how these are to be resolved.

Implementing the change

(1) Have the objectives been clarified well enough for everyone to understand?
(2) Will the change, once achieved, meet expectations aroused in the organisation?
(3) Is the responsibility for the change programme being placed at a sufficiently senior level?
   Have all the key tasks been identified and the person responsible for implementing them chosen?
   Is a timetable set?
(4) Are those responsible for introducing the change acceptable to others?
(5) How can I obtain continuing commitment from those involved in the change?
(6) How can I maintain momentum in the change programme?
(7) If time and energy are to be devoted to achieving the change, have I made arrangements so that my normal work activities are not neglected during the period of implementing the change?
(8) What are the means to be used to ensure that the new way of doing things is maintained?

Checking and monitoring change

(1) What quantitative and qualitative data am I collecting to show whether the change is achieving its objective?

(2) Has this change indicated the need for other changes?

Implementing change depends very much on the personal style of the manager but managers who successfully achieve change tend to have certain characteristics. These include:

- Interest in their own personal development, and the growth of their organisations.
- Enthusiasm for change.
- Desire for new experiences.
- Willingness to take reasonable risks.
- Openness to more than one course of action.
- Concern for achievement and interesting work.
- A somewhat unconventional approach to life.
- Willingness to accept others as experts.
- More inclined to examine current evidence than rely on past experience.
- Concern for human relations as well as efficiency.
- A planning and problem-solving approach.
- Open, participative relationships with other people.

There are a number of techniques which can be considered in implementing change:

(a) *Unfreeze – Change – Freeze*

Existing ways of doing things are often firmly established, and it is necessary to 'unfreeze' them by discussing them openly, critically examining them, experimenting with new approaches, and seeking new ideas from outside the situation. In this way, weaknesses in current arrangements can be more widely recognised, and people involved will be more inclined to accept changes. As change is introduced, new policies and procedures are made and these are then 'frozen' to become part of the new way of doing things.

(b) *Change agent*

Usually it is the 'manager' who is the 'change agent' but often it can be useful to involve an outside person or agency to help bring about change. For example, if a new approach to family planning is being introduced, it can be helpful to bring in the expertise of someone who is regularly involved in the new approach and knows what problems may be encountered. Such a person could be very helpful in giving advice, and helping local workers to understand what is involved, but the local manager cannot leave the whole task of introducing

Table 5.20 Force Field Analysis

All given situations can be seen as being in temporary equilibrium, with forces acting to change the situation (*driving* forces) being balanced by forces acting to resist the change (*restraining* forces). Examples of such forces are:

Organisational tradition	Initiatives
Current policies	Public demand
Resources	Distrust
Effect on status	Ambition
Organisational systems	Professional aspirations
Political pressure	Legal considerations

Force Field Analysis attempts to identify the forces and seeks to change their direction (driving or restraining) and/or strength (low; medium; high).

---

change to an outsider – the local manager still retains responsibility for managing the change.

(c) *A planning approach to management*

A management team which adopts a planning approach to its work will be committed to 'change' as a normal part of its work. Aims and objectives will be regularly reviewed, problems identified, solutions proposed, and action plans worked out. As action plans are regularly used and put into practice, it will be recognised that change is a normal feature of working life.

(d) *Force Field Analysis* (see table 5.20)

An analysis of what is involved in a change by specifying clearly the change which is to be made, the people who will be involved, and other significant factors. A list is then made of all those factors which are *for* the change (the helping factors), and another list of the factors which are *against* the change (the restraining factors). A manager's task is then to encourage and develop the factors which are operating in favour of the change, and to find ways of counteracting or minimising the factors which work against the change.

Table 5.21 is an analysis of what is involved in establishing a sub-centre to serve a population of 10 000 people.

After the Force Field Analysis of impeding and supporting factors, a check-list for organisational change can be helpful when introducing a specific change.

(e) *Preparing a 'domainal map'*

Reference was made to this useful managerial tool on page 166. When several key people are affected by a proposed change, a domainal map helps to

Table 5.21 **Force Field Analysis for establishing a sub-centre (A systematic approach to looking before we leap! Forces impeding and favouring the objective)**

Problem	Establishing a Level B station to serve 10 000 people

Step I	Specification of personnel, selection of village, accommodation, transport
	How many staff, what sort of staff (criteria)?
	Who will provide accommodation (for the facility and staff housing)?
	Criteria for selection of village (size, accessibility, water, communications, health problems of the catchment area, etc.)
	Type of transport.
	Finance.

Step II People involved

People	Role	Relation to District Health Management Team (DHMT)
Village Dev. Committee and Chief	Finding accommodation. Permission to start.	Link between DHMT and community
District Health Management Team	Supervision, training	–
District Education + Agriculture + Water and Social Welfare Officers	Support, advice and co-operation	Link to local government
Reg. Med. Officer and Nursing Officer	Provision of Level B workers, finance	Link to Ministry
Planning Unit	Training, Level C	Training, planning, advice

Step III Other significant factors

Present post, Level B worker
Personal interest toward work
Attractiveness of post
Basic and social amenities
Interest of community toward Level B

Step IV Restraining and helping factors

Restraining factors	*Helping factors*
Conflicts in Level B between curative and supervisory (outreach tasks)	Wish for professional health workers in the village
Cultural obstacles (such as objection to bicycles)	Working in a team is more attractive than working alone
Different expectations between community, Level B and DHMT	Living near work place
Lack of basic and social amenities	Personal contacts can give high job satisfaction

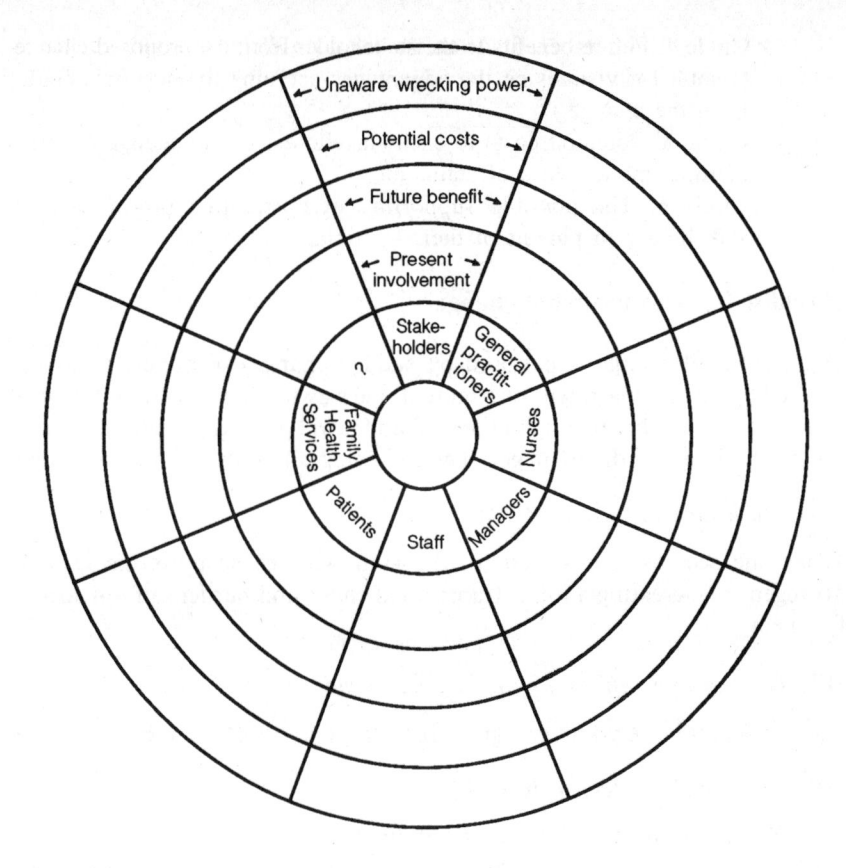

Figure 5.3 *Domainal map*

organise the information concerning each one of them for effective planning. Its usefulness lies in providing a visual display of each person's perceptions of the change, its benefits and costs, and exploration of each person's power to help or obstruct the change. When benefits outweigh the costs there is support. Obstruction occurs when the reverse is more likely.

A domainal map is created as follows (figure 5.3):

(1) Draw six concentric circles.
(2) Divide these circles into as many segments as there are stakeholders.
(3) Enter details in each circle:
   - Circle 1: Title of proposed change.
   - Circle 2: Names of stakeholders.
   - Circle 3: Present functions of the stakeholder in the situation where change is proposed. (Important for reflecting on the agenda of each stakeholder.)

- Circle 4: Future benefits to the stakeholder from the proposed change. (Useful for visualising the advantages accruing to each stakeholder from the change.)
- Circle 5: Potential costs to each stakeholder of the change in terms of time, money, loss of status etc.
- Circle 6: The possible supporting or obstructing power of each stakeholder in pursuit of their legitimate interests.

Checklist for organisational change

Organisational change is imposed by sudden changes of policy, or slowly evolved by one's department or begun at one's own initiative. But however it arises it can seem threatening to the staff and it will place great responsibilities on the shoulders of the manager. The following questions need to be raised:

How will it look to the staff?

What one sees as a new system, the staff will see as upsetting existing arrangements, creating a lot of transitional chaos and burdens of re-learning for them.

Who is going to be affected by the change – and who else?

The district is a closely-knit system and changes may affect a number of people.

How is it going to affect their work?

- Give them more *or* less responsibility?
- Make them feel more *or* less of a contributor to the whole District?
- Give them a greater *or* smaller sense of achievement?
- Make jobs more interesting and complete *or* more fragmentary and unrelated?
- Make existing skills more *or* less useful?
- Place heavy demands *or* no demands on learning new skills?

How is it going to affect relationships?

- Will new groups be formed and existing ones dissolved?
- Will any individual become or feel isolated?
- Will some good jobs be created which change the status structure of the District?
- Will it affect people's lives outside the work situation?

Has enough time been allowed for the following?

- Discussion with other managers?

- Discussion with staff?
- Replanning the whole thing if one needs to in the light of these discussions?

Is a 'third party' involved?

Occasionally consultants and specialists assist such changes. One should try:

- To understand their terms of reference clearly.
- To ensure that these are known and explained and accepted by the staff involved.
- To set up working teams with the specialists – not isolate them.
- To remember that their intrusion is commonly resented and one must try not to let such resentment influence one's behaviour.

MANAGING CONFLICTS

In the management of health care conflicts and differences of opinion are inevitable. To give just a few examples – because money is short, health managers will frequently be in discussion with government or other funding agencies to obtain more, and will be in competition with other departments for funds. There will be differences of view within Health Districts on priorities – which diseases should be dealth with first, which village should have more facilities. Workers themselves will have differences – over their pay and conditions of work, working arrangements, and so on. In the Health Team itself there may well be differences on such things as the importance to be given to preventive or curative medicine and the roles of health workers, for example, nurses, medical assistants. The Health Team which is committed to rapid development of Primary Health Care is likely to find that a good deal of its time is spent in dealing with conflicts and reconciling different points of view.

A number of methods may be utilised to manage conflicts (see table 5.22) and as follows:

The problem-solving approach

The basis of this approach is that the best solution to a problem is one that is agreed by all the parties who are affected by the problem. It only works, however, where there is common agreement on the fundamental aims of the group. For example, if there is disagreement in a Health Centre over the allocation of rooms and times for different clinics, those involved can probably discuss the issue and resolve it provided there is general agreement on the overall aims and priorities for the Health Centre. Where there is fundamental

Table 5.22 **Handling conflict**

	CO-OPERATIVE
ACCOMODATING: neglecting one's own concerns to satisfy those of the other party e.g. • selfless and generous • accepting an instruction when one would prefer not to • yielding to another point of view	COLLABORATING: working with the other party to find some solution which fully satisfies both parties e.g. • exploring disagreements • resolving a situation which would otherwise lead to competition for resources • finding a creative solution to an interpersonal problem

UNASSERTIVE ⟵ COMPROMISING: finding an expedient, mutually partially satisfying to both parties e.g.
• splitting the difference
• exchanging concessions ⟶ ASSERTIVE

AVOIDING: not pursuing one's own concerns or those of the other party. Not addressing the conflict e.g. • side stepping an issue • postponing an issue • withdrawing from confrontation	COMPETING: pursuing one's own concerns at the expense of the other party using ability to argue, status, economic sanctions e.g. • standing up for own rights • defending a position
	UNCO-OPERATIVE

disagreement on the policy for the Health Centre, however, it is unlikely that any amount of discussion will provide a solution. This is why it is important to have agreed aims which are accepted and well understood throughout a Health District. Given such aims, the vast majority of problems can be dealt with through rational discussion and argument. In a problem-solving approach the group concerned goes through the steps of problem solving just as a manager would if he were solving the problems on his own, namely: defining the problem clearly, setting clear objectives for solving the problem, analysing the problem to determine causes and effects, searching for alternative solutions,

Table 5.23 **A framework for problem analysis**

Who or what has the problem?	Who or what does not have the problem?	What is distinctive about the difference?
Where is the problem located?	Where is the problem not located?	What is distinctive about the difference?
When does the problem occur?	When does the problem not occur?	What is distinctive about the difference?
How big is the problem?	How big or not is the problem?	What is distinctive about the difference?

deciding on the solution which best meets the objectives and agreeing the specific action to be taken (see table 5.23).

The benefits of a problem-solving approach are that it can strengthen the involvement and commitment of all those concerned, it can channel people's energies away from destructive conflict towards a constructive search for solutions and it can be a valuable way of generating new ideas and approaches.

Bargaining

This takes place when two parties who are in dispute do not share common aims. There is a large element of bargaining involved in local purchasing, and local personnel may be highly skilled and enjoy the processes of making offers, counter-offers and 'walk-aways' to obtain goods at the lowest possible prices. At a higher organisational level, hard bargaining is called for in deciding expenditure between hospital care and community health. Another example is that of negotiations with drug companies and other suppliers to ensure the supply of the right goods, at the right time, in the right place and at a reasonable cost.

Use of 'third parties'

Where two parties are in dispute, a third party is sometimes called in to settle the issue. Frequently a manager is put in this position when two or more members of staff are unable to agree on an important matter. The manager has the managerial authority to decide one way or the other, according to what is best for the District, and often that decision will be accepted by the parties concerned. But there are times when it is better for the manager not to *arbitrate* and make the decision, but to *conciliate*, that is, to bring the disputing parties together and guide the discussion so that the parties are forced to sort the problem out for themselves.

Avoiding or ignoring the conflict

Managers may pretend that conflict does not exist because they are not sure how to deal with it, or are afraid the conflict may get worse if they get too involved. Unfortunately, unresolved conflicts often tend to get worse. Managers therefore need to be constantly 'listening' for the first signs of major difficulties so that conflicts can be dealt with sooner rather than later. It helps to recognise that differences of opinion and interest are an inevitable part of the management task. Effective managers are skilled in dealing with differences and recognise some of their positive side-effects. Differences can be used as a means of generating new ideas and approaches, through the use of problem-solving discussions to resolve difficulties. Through facing up to the realities of power and individual interest, managers learn how to advance the cause of improved health for the community, and how to win the support of powerful interest groups. And by successfully resolving conflicts, managers are able to build up strong relationships with both health workers and others in the community.

Confrontation

If confrontation is handled well it does not result in conflict. But good confrontation calls for good interpersonal skills. Making one's feelings and needs known while at the same time recognising and respecting the other person's is the art of constructive confrontation. It helps to bear in mind that confrontation is always a shock to the other person. Hence there is a need to support the person. Confront the issue and challenge the pattern but never the person.

Before deciding on a confrontation there are three things to consider: (a) the issue; (b) the behaviour and (c) the source of the irritation.

The *issues* may be trivial, but feelings may be running high. If emotions aroused are far in excess of the issues it indicates some cause from the past. If problems were not dealt with then, they come to haunt future interactions and the earlier the issue is aired the better.

People's *behaviour* is not easy to deal with, because it is part of their personality. On the other hand, if one can get to the *source* solutions may be easier to find.

During confrontation the manager may adopt one of the following roles to suit the occasion, bearing in mind that overdoing any of the roles creates an adverse impression:

Roles	Behaviour
Controlling parent	Being firm
Too much of it	One is considered bossy and overbearing. It can cause grievance and even open rebellion.
Nurturing parent	When it is necessary to care for someone.
Too much of it	One is considered overfussy and smothering. Causing dependence and denying to others the opportunity to develop.
Adult	Being logical and rational. Appearing to be deciding between conflicting demands.
Too much of it	Accused of being a 'robot'.
Adoptive parent	Full of politeness and well behaved.
To much of it	Considered passive and unable to resist unreasonable demands.
Colleague	When one needs to let others know how one feels. Expressing fun, pleasure or genuine anger and disappointment.

GIVING SUPPORT TO SUPERVISORS

The job of a supervisor is both responsible and difficult. Supervisors frequently have had little formal training in supervision, work in geographical isolation, and tackle all the problems of organising work and supervising staff in the most difficult circumstances. Managers who can give good support to supervisors will find that it results in the improved quality of health services.

Generally, supervisors become so through promotion, because they have been good workers themselves. This can be a great strength in that the supervisor knows thoroughly the work which has to be done, the practical problems and how to get over them. The supervisor is also likely to have a close affinity with other workers, often coming from the same social background, speaking the same language, and sharing a common approach to life. But ability to do a job well does not automatically make someone a good supervisor of other people.

The job of supervisors

There are certain elements which are frequently found in the work of supervisors, which include the following:
(1) To plan the work of the department/section/health facility.
(2) To allocate work to individuals.
(3) To co-ordinate the work of different people.

(4) To see that work is done to a proper standard.
(5) To communicate
   (a) to workers – aims and objectives, work to be done, changes, and so on;
   (b) to managers – reports, needs, difficulties.
(6) To demonstrate, train, support, help and encourage workers to do their work well.
(7) To solve workers' problems when needed.
(8) To improve working methods.
(9) To maintain healthy and safe premises and working methods.
(10) To ensure a good working environment – physically and socially.
(11) To act as the level of policy-making closest to workers.

The supervisor should be well aware that to do the job well certain supervisory 'tools' are needed. These are akin to 'tools of the trade' which a craftsman or other workers would use and include such things as:

Schedules/Timetables/Diaries/Programmes – because much of a supervisor's work consists of getting certain things done at certain times (for example, clinic sessions, supplies requisitions, visiting programmes, and so on), a systematic way of planning and controlling such activities is needed.
Instruction guides and procedures – to help with work which is of a semi-routine or systematic nature, for example, procedures for weighing and examining under-fives, diagnosing and treating common complaints, reviewing activities of the village health workers and so on.
Rules and regulations – these should be simple and clear to understand. Out-of-date and complicated regulations can be more of a hindrance than a help.
Budgets – this refers usually to money which can be spent within a set period of time. For the supervisor a more tangible set of budgets is needed, for example, numbers of staff and working hours available for certain tasks; amounts of overtime or extra payments which may be authorised; amounts of drugs, vaccines and other supplies available per month; mileage and travelling allowances per month. These are budgets which a supervisor should account for in some detail and exercise close control over.

Support for supervisors

Above all supervisors need support from their own 'supervisors' or managers and this should include:

Leadership in the form of clear guidance from their own manager when needed; trust, and regular contact.
Training in preparation and on appointment as a supervisor; and as a

continuing process, both on and off the job. Because of the practical nature of the work, supervisors can gain much by discussing with other supervisors various aspects of their jobs.

Backing for decisions. Supervisors must know that they will get support whenever they have to make decisions within their own sphere of responsibility. Even where bad decisions are made, it is important that supervisors receive support especially in the eyes of their own staff, whilst being given help to redress the situation and guidance on how to make better decisions in the future. It is better for the occasional mistake to be made than for supervisors to be afraid of making any decision.

Recognition and proper status for the supervisor's own position. This means that a manager should not try to 'by-pass' the supervisor in dealing with members of staff, and as much as possible should work with the supervisor in explaining and discussing future plans, changes, and so on, with staff.

Reliability. Despite the many problems involved, supervisors should be able to rely on supplies of drugs and other materials arriving when expected; a regular pattern of visits from their managers; and above all action promised by managers should in fact be taken.

Involvement in planning. Because supervisors are nearest to the 'grass roots' of the health service, they have an important contribution to make in planning for the future. Through sharing in planning, supervisors can develop a broader outlook on the task of health provision and further their own personal development to take on increased managerial responsibility in the future.

PERSONNEL – THE MANAGEMENT OF HEALTH WORKERS

Rural health services rely on the efficiency and enthusiasm of the staff who run them. 'Personnel Management' is the aspect of management which ensures that everything concerned with the employment of staff within the Health District is done effectively, so that staff can be happy in their work and give of their best. Much of the job of personnel management is done by managers within the District such as the District Medical Officer, Head of Nursing, Heads of Health Centres and so on. But other important aspects of personnel management (for example, establishment of personnel policies and procedures, negotiation and payment of salaries and wages, the running of training schemes) may be carried out centrally, at District, Regional or National level. Many personnel practices may be already established in a well-run hospital, and the task of the District Medical Officer may be to ensure that these practices apply within the rural sector of the District. In other cases existing personnel policies and practices may need to be adapted for the Primary Health Sector.

The management of health workers includes:

Recruitment and selection – searching for and choosing workers for particular posts. By taking the trouble to select good workers, many problems can be avoided and a high quality of work achieved.

Induction and training – ensuring that new workers are properly introduced to their work, and continue to be trained to do their work well.

Allocating work – giving workers tasks which are within their capabilities and make the best use of their skills and knowledge.

Supervision – ensuring that workers can always get help when they need it and that high standards of work are maintained.

Discipline and grievance – Dealing effectively but fairly with workers who break rules or whose work is not up to standard; and settling speedily any grievances or disputes which workers themselves may have.

Good communications – ensuring that there is two-way open communication between workers, their supervisors and members of the Health Team.

Counselling and guidance – helping workers to solve day-to-day difficulties, or personal problems which affect their work.

Promotion and career development – giving workers the chance to 'better themselves' whilst ensuring a supply of workers who can take on more responsibility in the future.

Manpower planning

Manpower planning is concerned with 'team-building' for the future. Managers need to consider future staffing needs, existing and potential sources of supply, recruitment, training and development strategies to ensure that trained staff will be available to do the work called for in the District's long-term plans.

In most countries recruitment and selection of staff occurs centrally at the National level and the District Medical Officer will find his staff appointed by a distant Ministry of Health. Many of them may be recent graduates with little experience. Here the challenge is to mould the various workers into a team and to identify from amongst the team members those who show promise for future promotion to positions of responsibility. As an aid to selection a technique which has been widely adopted is that of the 'seven-point plan'. This is a means of considering the factors which should appear in a personnel specification under seven main headings, as follows:

Physical make-up

Are there any important defects of health or physique? How important are appearance, bearing and speech?

Attainments

What level of education and training, occupational training and experience is needed?

General intelligence

What level of intelligence is needed? Intelligence is not always related to education!

Special aptitudes

Is there a need for any special skills? Facility in the use of words? Or figures? Any unusual talents?

Interests

Are these intellectual? Practical? Physical? Social? Artistic? and so on.

Disposition

For example, acceptability to other people. Ability to influence others? Dependability. Self-reliance.

Circumstances

Family responsibilities, housing needs, and so on. What do other members of the family do for a living? Are there any special openings available?

The aim of the Seven-Point Plan is to try to obtain as complete and objective a picture of the candidate as possible. It can reduce the subjective element in selection, and act as a frame of reference in comparing one person with another. It also provides a systematic framework for asking questions of candidates and carrying out an effective interview.

In appointing staff to various positions it will help to have a full description of the job the individual is expected to do. This will help in 'fitting' the most suitable person to a job. Moreover, the process of writing a job description entails job analysis which provides an insight into what is expected and required in the job. In turn this helps with job evaluation, work organisation and review of the candidate's performance. The exercise may also raise questions of training, and any deficiency can be rectified through in-service training programmes. These various inter-relationships are shown diagramatically in figure 4.8 (page 154).

TRAINING

Rural health services largely depend on auxiliary health personnel. The rural health organisation has been described as a 'skill pyramid' in which skills in health care are widely distributed amongst those who provide health services, and not restricted to doctors and nurses who have had intensive medical and clinical training. Training needs to be 'built in' as part of the normal activity of the District, and not seen solely as something which takes place in training schools or is 'added on' to the normal working routine when there is time to do it.

There needs to be a *present* and a *future* orientation to the way in which people's work is viewed. Staff need training to carry out their existing work competently but they also need to learn how to achieve higher levels of competence in the future, either within their existing work, or by promotion to a job with greater responsibilities. There needs then to be a *training plan* for every worker in the Health District. It is unlikely that this will always be written down (though there is merit in writing it down annually for each *senior* member of staff), but each worker and his supervisor should have a clear idea of which aspects of the work they are currently learning to improve and which new skills they are developing. Ideally every team leader or supervisor should have a Training Plan for their own unit indicating who needs to learn what and how urgently. The most important Training Plan in the District is that of the District Medical Officer himself. Table 5.24 indicates what such a Training Plan might look like, and obviously would need to be adapted to local circumstances, but it can give managers some basis for organising training.

It is clear that the type of training suggested is not of the 'recent advances' or 'update' kind, but practical training related to the day-to-day work of the individual. Besides the training plan, resources need to be made available. Time and money spent on training is not a misuse of resources but can lead to large 'pay-offs' in the future. Health workers need to see their work as a training task – the population being taught how to maintain its own health, and workers learning how to carry out increased responsibilities. If this attitude is prevalent then many working situations can be used as training opportunities.

Staff training will fall into the following categories:

- New health workers – induction training to acquaint them with their responsibilities and local circumstances.
- Staff about to take on new responsibilities.

Table 5.24 Example of a training plan

Training plan Dr............. D.M.O............. District.............

Training plan	Supervisory Skills							'Technical' Skills						Management Skills					'Broadening' Skills			
	Giving instructions and training	Dealing with staff's problems	Communication	Allocating and organising work	Dealing with sub-standard work	Support to village health workers	Team leadership	Drug prescribing	Referrals	Health education	Treatment	Diagnosis	Information for evaluation	Financial control	Managing drugs & supplies	Forward planning	Improved clinical knowledge	Wider management skills	Understanding of community	Other aspects of health care
Mr A. Health Centre Supervisor	B	B	A	A	B	B	C													
Mrs B. Health Centre Head Nurse	B	A	A	C	A	A	B													
Mrs C Health Centre Public Health Nurse	C	B	B	B	A	A	A													
Miss D. Community Midwife	A	C	A	B	A	B	B	MCH Organisation A B B												
Mr E. Chief Sanitarian								Appropriate technology for: (a) water — A (b) sanitation — B												

Key: C = Competent
B = Needs improvement
A = Needs major improvement

- 'Students' (for example, trainee community nurses) attached for practical training.
- Those whose work is not up to standard.
- Those who need refresher training, for example, community health workers, traditional birth attendants, and so on.

STAFF DEVELOPMENT

Just as the District Health Management makes plans to improve the health services, so also should they plan the career development of workers. Individuals should be helped to improve their skills and qualifications, and move on to posts of greater responsibility. Many bright young people who get some education and training in the District area leave for the towns because they see the urban market as the only way of realising their ambitions. It should be possible to retain a number of these people if the District Health System offers opportunities for advancement.

Staff development is also essential for continuing improvements in the quality of care provided in the District. Education, experience and culture are recognised as the triangle of forces for the development of the individual. The shortcomings of medical and nursing education are well known. Recent graduates are singularly unable to manage the uncertainty of diagnoses, and judging response to treatment in situations where back-up of any kind is not easily available. On the other hand, experience needs to be 'internalised' to be of any value. This is best done in educational discussions, meetings and workshops. But education and experience are being interpreted by the individual in the light of his own background of values and beliefs i.e. culture. Values and beliefs continue to be inculcated as we clamber up the ladder. All these interacting influences have to be taken into account in planning staff development, and when considering the following:

(1) *Selection policies.* As far as possible workers should come from the village or District being served. Workers should be acceptable to the local community, but those with education and initiative should be encouraged to take on more responsibility in the future.
(2) *Promotion.* First consideration should be given to people already working in the health district. This does not mean appointing people who are not properly qualified, but that training and developing of staff is a major responsibility of the District Health Manager.
(3) *Future opportunities.* There should be realistic opportunities for those with initiative, and workers should be able to see how they can progress to more responsible work. Lack of basic education will be a bar to many positions, and opportunities need to be created for the training and advancement of those with little or no basic education.

(4) *Links with training institutions* both within and outside the District. Qualified nurses are likely to be trained at a Regional or 'Teaching' hospital, medical assistants may be trained at a Regional Centre. An 'active' District will be looking for candidates for these posts from the District and the local community, encouraging the training schools to use the District for field attachments, providing 'vacation' work opportunities where appropriate, and encouraging students to return to the District after training. It could be the personal responsibility of a District Medical Officer to seek out and maintain links with medical students who have been 'on attachment' or spent 'elective periods' in the District and who might ultimately return to work as an assistant!
(5) *Good internal training in the District.* To ensure competent workers who are able to take on more responsibility in the future.
(6) *Regular appraisal.* The opportunity for more senior staff (for example, Health Centre Supervisors) to meet with their senior manager specifically to talk about their own work performance, how it might be improved, and what can be done about their own personal development.
(7) *Opportunities at the District level of health care.* There may be openings for Primary Health Care staff in the District Hospital, which can help hospital staff to have a better understanding of the needs of the wider community.
(8) *Flexibility.* A narrow 'professionalism' has constantly to be fought. For example, doctors, nurses, sanitary inspectors or health assistants jealously guard their own particular expertise and refuse to share it with others, or allow others to do work which traditionally they have regarded as their own. The more a number of workers exist who can step into another's role, the more each worker benefits and is equipped for broader responsibilities in the future.
(9) *Support for educational opportunities.* Close links with primary and secondary schools can ensure that Health Education and health matters are dealt with in the schools, but also can ensure that workers get help with their own education. Active involvement in such activities as the 'Child to Child' programme, adult education classes, and distance teaching by radio programmes can lead to continuing improvement in general education for health workers.
(10) *Balance between educated and experienced workers.* As educational opportunities improve there are dangers that better educated, younger people come in to take all the interesting, responsible work and block off opportunities for hard working, conscientious but less well-educated workers who normally advance through the ranks by accumulating experience. It is very important to be sensitive to the problems this can cause and ensure that there is a proper balance between workers and opportunities to which all can aspire.

(11) *Coaching*. This implies the conscious, systematic and regular effort to learn from normal work activities. People often learn best from what they find out for themselves. So a District Medical Officer on regular visits to a Health Centre Supervisor can guide, explain and assign more challenging tasks from which the Supervisor can 'learn by doing'.

(12) *Local study days*. There may be transport problems to be overcome, but the effort may be well worthwhile of arranging for health workers from right across a District to come together to listen to a special speaker, for example, a Regional Agricultural Officer, or Regional Medical Officer, or to see some new technique, simple equipment or teaching aid. Such a day also provides the opportunity to consider how new ideas and approaches can be implemented throughout a District.

(13) *Distance learning*. This method of teaching/learning is gaining popularity, ever since the success of entire universities based on such learning. A large number of sets of material are already available, and new ones are continually being developed. This approach to learning entails working through material in the form of tutorial texts, followed by an assignment to prepare and classroom teaching at regular intervals.

(14) *Recognition of adult learning psychology*. Adult learning is different because adults have experiences which they need to link up with the new topics they learn. They learn better what is relevant, especially if it draws on their reasoning and observations rather than learning by rote (table 5.25).

The above suggestions indicate a wide range of activities which can improve the development of staff in a District. There may not be time to engage in all of them simultaneously, but the key is to start on some of them. Staff development activities can require a lot of attention in the first instance, but once under way, can grow and multiply with workers taking more and more initiative for their own development. The fundamental requirement is an attitude of mind that development of staff is essential for better performance and improvement of services.

MAINTAINING STANDARDS AND DISCIPLINE

Those in managerial and leadership positions need to have high standards for their own work, and maintain similar high standards throughout the organisation. Where high standards are set and achieved, and where workers know what standard of work is expected of them, it should rarely be necessary to take disciplinary action against any worker provided care has been taken in the selection and training of the workers and they are given support and supervision. However, there are occasions when someone's work does not come up to expectation and disciplinary action has to be taken. For such occasions disciplinary rules and procedures are needed which are understood, and accepted as fair and reasonable.

Table 5.25 **When should particular learning methods be used?**

Method	Description/examples	Appropriate uses	Inappropriate uses
Formal lectures	Teacher 'delivers' information to students who are expected to learn it	Transmission of knowledge, procedures, or facts that have to be memorised; also good for large audiences	To teach skills or problem-solving techniques
Small group discussions	Trainees discuss problems and possible solutions; teacher stimulates discussion but does not dominate	To develop problem-solving skills; to provide feedback to instructors	To teach facts that must be memorised
Practical experience	Students practice clinical and community skills, generally after prior demonstration or instruction	For learning specific technical functions	To teach facts that must be memorised
Apprenticeships	Students assist an established practitioner and learn by observation, imitation, and practice	For learning clinical skills, especially highly technical ones	For basic learning (apprentices should have some skills already); for teaching community work
Role-playing	Some students 'act out' typical work problems; others observe, subsequently discuss	To give students practice in patient and community relations	To teach students treatment requirements with which they are not already familiar
Storytelling	Trainer tells a close-to-local life scenario illustrating causes/consequences of ill-health; discussion follows to draw out 'lessons'	To demonstrate how social conditions and personal behaviour influence health	To teach technical diagnostic and treatment skills

Table 5.25 (contd.)

Method	Description/examples	Appropriate uses	Inappropriate uses
Solving real problems	Students develop solutions of their own with some coaching; groups critique/refine these solutions	On-the-job and inservice training	For work in totally unfamiliar subject areas
Village theatre	A combination of role-playing and storytelling, usually using traditional theatrical medium	For teaching the social and personal causes of ill-health to large groups	For teaching technical skills

Method	Advantages/strengths	Disadvantages/weaknesses	Comments
Formal lectures	Easy for trainer; covers much material in short time	Students may not learn well; teacher does not learn about field problems or perceptions of students	Widely misused
Small group discussions	Encourages students to discuss things on their own; helps instructor to understand what students are thinking; helps develop attitudes	May 'drift' from planned agenda; trainees may not learn what trainers want them to; can be time-consuming	Discussions occur whether planned or not; formal training structure may discourage them
Practical experience	The only effective way for perfecting skills; also develops attitudes	Requires close supervision and persons or objects on which to practice	Most training programmes under-emphasise practice, partly because training sites do not represent typical working environments

Apprenticeship	Facilitates practical experience under close supervision	Preceptors often lack experience/interest in supervision and teaching; may use apprentices inappropriately	Apprenticeship arrangements usually very different from eventual CHW working environment
Role-playing	Develops problem-solving skills and attitudes	Students often have difficulty putting themselves in unfamiliar roles	Can be a waste of time if not well done
Storytelling	Helps trainees draw lessons from subsequent 'real world' situations and provides community health education	Trainer (or a coached trainee) must have storytelling skills; only effective if followed by good group discussion	Closely modelled after real-world learning environment
Solving new problems	CHWs often know problems and the feasibility of potential solutions better than instructors	Trial and error involves some errors, and these may sometimes be costly	Sharing of problems and solutions among experienced CHWs can be far more useful than formal training
Village theatre	People may learn best from a traditional medium	Often requires considerable preparation	Usually used for generalised health education rather than for CHW training

Disciplinary rules should define the minimum standards of behaviour at work like, for example, work performance standards, requirements to obey reasonable instructions and the prescribed punishment for serious misconduct.

A disciplinary procedure sets out action to be taken when rules are broken or work is unsatisfactory. Procedures enable managers to deal effectively with the situation, and deal fairly with the worker concerned. Disciplinary procedures should:

(1) Be in writing.
(2) Specify to which grades of staff they apply (for example, there may be different procedures for professional staff and less skilled workers).
(3) Provide for speedy operation.
(4) State the range of disciplinary actions that can be taken (for example, dismissal, suspension with or without pay, reprimand, transfer).
(5) Specify those who have authority to take action (for example, a procedure may state that a Health Centre Supervisor has authority to reprimand but not to dismiss).
(6) Provide for investigation before disciplinary action is taken. Where the rules provide for instant dismissal (for example, for serious misconduct) the first step may need to be suspension while the case is investigated. An action taken in the heat of the moment may cause difficulties later.
(7) Ensure that individuals are informed of any complaint and given opportunity to state their case.
(8) Give an individual the right to be accompanied by a representative or friend.
(9) Ensure that reasons are given for any penalty imposed.
(10) Give a right of appeal if individuals feel they have been unfairly treated. This appeal may be to a higher level of management, or where appropriate to a village committee or a local leader in the case of a village health worker.
(11) Give individuals the opportunity to improve their conduct after a first warning.
(12) Provide for one or more warnings (depending on the seriousness of the offence).
(13) Ensure that authority to dismiss does not rest with a worker's immediate supervisor, but with a more senior person. Dismissal should normally only occur where repeated warnings have been ignored or there is gross misconduct.
(14) Provide for dismissal if an offence is repeated after a final warning.
(15) Ensure that all workers know what rules and procedures apply to them, and that managers and supervisors also clearly understand the part they have to play.

In a number of countries employment legislation requires employers to deal reasonably with employees in disciplinary matters. Such legislation should not make it harder to discipline or dismiss unsatisfactory workers. With clearly understood rules and procedures both managers and workers know where they stand, and if disciplinary action is needed, it can be dealt with in a reasonable and straightforward manner. Even with well-understood rules and procedures, the most important factors in maintaining high standards of work and discipline are the standards which managers and supervisors set for themselves, and the quality of support and supervision they give to their workers.

COUNSELLING – OR HELPING STAFF WITH THEIR PROBLEMS

'Counselling' is an important part of the job for many health workers. For example, a public health nurse will 'counsel' or discuss with a mother those personal and family problems which affect the growth and development of her baby. But health workers themselves have problems of their own which may be personal or related to work, and managers are often asked for help and advice.

So much of the Health Manager's task consists of supporting, guiding, and helping other people to do things well that it is worth looking at some of the helping, 'counselling' relationships. The danger is that managers often feel that they know best how to offer solutions whereas what is important is that workers need to be helped to define their own problems, clarify and explore them, and try to find realistic solutions. It is obvious that managers need to be good at listening and at encouraging workers with a problem to express their feelings about it. This requires a high degree of concentration and the ability to observe what the worker is trying to say. Communication between individuals can be difficult because of personal, psychological and sociological differences. These can be accentuated when the manager has a different regional, ethnic and language background, a higher 'professional' status, and little knowledge of the details of someone's day-to-day life and work.

Some managers feel happier with a 'directive' and others with a 'non-directive' approach to counselling. In the directive approach the manager will give clear direction and advice. In non-directive counselling the manager does not suggest solutions but helps workers solve their own problems. There are situations which call for a 'directive' approach, for example where standards of work have to be maintained; but the effective manager is the one who can adopt a 'directive' or 'non-directive' approach depending on the circumstances.

FINANCE

Managers need to ensure that Primary Health Care actually gets all the financial support that is available to it and that money does not get siphoned off into hospitals; is used effectively and is not wasted either fraudulently or through inefficiency, and is available for future as well as present needs. To meet such needs various financial techniques and systems have been developed. The best systems are often the simplest, but the important thing is that the Health Management Team must work closely with their financial colleagues, share an understanding of each other's functions, and develop practical systems which effectively meet their needs.

Although the development of Primary Health Care (PHC) may be stipulated in a District Development Plan, the implementation of that plan will be highly dependent on the finance available. If for any reason, for example, inflation, economic recession, political changes, those allocations are reduced, there will be a danger of the PHC programme being cut back to ensure the continuance of hospital services. Moreover, there is a natural tendency for the financial demands of the hospital to grow. Spending patterns and proposals in the hospital should therefore be critically examined to ensure that they do not lead to further imbalances between primary and secondary care.

Sources of finance

Sources of finance should be clearly identified, and the best possible case put forward for obtaining a satisfactory allocation. Hospital services have often developed faster than primary care because hospital doctors and managers have been highly skilled at putting forward the case for finance to develop those services. As governments increasingly provide the main funds for health services, a clear understanding of how government funding arrangements operate is needed.

Other non-governmental sources of income should not be neglected. Grants from aid and development agencies, missions, and so on, can be important sources of finance, perhaps related to individual or special projects, often of an innovative or enterprising nature. The key to securing or maintaining such grants depends on the strength and clarity of the case put forward. The case should include (a) a specification of the *goals* to be achieved (for example, a nutrition project, immunisation programme, establishment of a health post), (b) details of the *action plan* (milestones) proposed to achieve the goals, (c) identification of the *resources* needed (for example, initial expenses, building, equipment, and so on, and operating expenses – staff, supplies, and so on), and (d) an outline *budget* showing how much finance is needed for each year of the project for each item of expense. Much of this information will be included in the District Plan. Moreover, help is usually available from aid agencies which normally provide details of what information is needed and how to present it.

An important part of the Primary Health Care concept is that as far as possible local communities should support their own health services. In some places fees for services, for example, attendances at Health Centres and medical consultations, are charged for those who can afford it; in others, villages may raise levies (for example, on agricultural produce at the time of sale) to help pay for health services, or may support village health workers in cash or in kind. Charges may also be made for supplies of drugs, nutritional supplements, contraceptive devices and so on. There is a delicate balance to be maintained in raising finance locally. On the one hand, it can be good for people to contribute to their own health services because a service that is paid for is often valued and appreciated and put to good use. There can also be a greater sense of involvement where money is raised locally and used on local projects, rather than raised through taxation and channelled back into health services through central government funding. On the other hand, when charges are made there is the very real danger that those who most need health services will not get them because they cannot afford them; also that the most beneficial and cost-effective services for example, preventive or environmental health services will not be financed because they are usually not felt to be important, and cannot be paid for through individual charges. Resentment may also build up where local people who are already paying high taxes without seeing any benefit to themselves are asked to finance their own health services. In addition, when local charges are made, the administrative costs of collection, coupled with the hazards of fraud and misappropriation, can outweigh the benefits of the funds actually raised.

Health Care Plans and Programmes need to be translated into *financial plans* concerned with estimated capital and revenue expenditure for both the longer-term Strategic Plan and the shorter-term Operational Plan. As capital expenditure is concerned with expensive items of buildings or equipment, the expenditure is often spread over a number of years and can be shown in the form of a *capital plan*.

A *revenue plan* relates to the day-to-day running of services. The financial plan should show the amounts allocated for the current year, as well as for future years. Estimates and forecasts for future years need to take into account inflation and proposed changes in the levels of services to be provided and income expected.

Operating budgets

Capital and revenue plans show the future plans for the District expressed in financial terms over a number of years. Operating budgets are concerned with the application of that plan for the current year. The first step is to obtain a more precise estimate of income and expenditure for the coming year. The predictions in the financial plan need to be reassessed in the light of income and expenditure in the previous year, and a new budget prepared.

It is a good principle of management that budgets should be held at as low a level in the organisation as practicable, so as to give those who actually use resources of money, staff, supplies, and so on, the responsibility for controlling them and using them effectively. The District Health Team should have overall control of the District Health Budget, with individual managers accountable for parts of it, for example, the District Nursing Officer controlling the nursing budget. Once a budgeting system is established and working well, greater responsibility for budget control can be given to those more closely involved in the day to day provision of health services. This is done by establishing 'budget centres', for example, individual Health Centres, a mobile team, primary health services based at the District Hospital, and so on. The appropriate 'manager', for example, the Health Centre Supervisor or leader of the mobile team as the case may be, is involved in preparing, controlling and managing that part of the budget, whilst members of the District Health Team retain over-all budget responsibility.

Budgets offer an effective means of planning, implementation and evaluation. When budgets are prepared with desired outcomes or utilities in view, the process can be referred to as programmed budgeting. Money is allocated to specific programmes within the District Health Plan, for example, to health education, family planning, immunisation, rather than to 'lines', that is, furniture, refrigerators, equipment and so on. Programme budgeting allows health programmes to be viewed as part of an over-all health plan designed to achieve defined and measurable objectives of the whole health plan of the District.

The allocation of funds to programmes must be done in such a way that the programmes likely to contribute more substantially to the realisation of the aims and objectives of the District Health Plans are weighted. This enables scarce funds to be distributed more rationally. For programme budgets to be costed accurately and uniformly, 'standard cost lists' provided for various economic zones of the whole country by an appropriate government agency can be very helpful. Such 'standard cost lists' contain information on salaries and allowances for professional and other health workers and costs of standard equipment, vehicles and spare parts, and so on.

Programme planning and budgeting is a useful technique because it compels District health staff, indeed all health staff, to plan carefully their health activities and set well-defined and measurable objectives. This way the District budget becomes a management tool which can provide the feedback (evaluation) necessary for the redefining of objectives should this be necessary.

Costing information

Relevant costing information can help a District Health Team in its management task. A number of questions come to mind: Which are the most expensive health centres to run in the District? Which are the least expensive? What is the cost of treating a particular condition with Drug A as opposed to Drug B? What is the cost of running Mother and Child Health Clinics in different parts of the District? What is the cost of running such clinics with fully trained as compared with less well-trained workers? What are the cost implications of alternative health care strategies?

Costing is the process of bringing together all the costs associated with a particular aspect of the health services (for example, a Health Centre and its outlying village services; a District's sanitation or environmental health programme; a training school for health auxiliaries; a specific immunisation programme) in order to assess what it actually costs to provide that service. Costing information is needed to assess the feasibility of new programmes or services; to assess whether services are becoming more or less expensive; to make cost comparisons between different programmes and facilities (for example, different Health Centres); and to provide control information for Government or other funding agencies.

Key elements in costing include:

(1) Identifying 'Cost Centres' – for example, Health Centre X, Malaria control, Vaccination programme, and so on.
(2) Allocating costs, including direct and indirect costs, to the Cost Centre
   (a) Direct costs – costs directly affecting the Cost Centre, for example, salary of Health Centre Superintendent, cost of malaria control.
   (b) Indirect costs – Costs 'shared' by the Cost Centre with other cost centres. For example, a proportion of the cost of the DMOH (salary, and so on) as part of the DMOH's time is spent in supporting the work of the Health Centre or programme.
(3) Quantification of 'units of service' provided for example, numbers of patients seen in a Health Centre or clinic, numbers of vaccinations given.
(4) Cost ratios. A combination of (b) and (c) which shows the cost of a particular unit of service, for example,

$$\frac{\text{Total costs of vaccination programme in one year}}{\text{Numbers of people vaccinated}} = \text{Cost per vaccination}$$

This simple 'cost per person vaccinated' can then be used to assess the value of a particular vaccination programme, to determine charges, or to make cost comparisons between different vaccination programmes.

Cost-benefit analysis

It is necessary to examine from time to time the trade-offs between the costs and benefits of various health interventions at the different service points in the District. Both costs and benefits are different from a community angle as compared to treatment in the individual patient. A given intervention may work well in the individual but a number of behavioural factors come into play when the same intervention is to be assessed at the community level, for example, oral rehydration therapy (ORT). In the hospital and clinic setting ORT will be used for severe cases of dehydration by better trained doctors and nurses. In the community setting, ORT may not be practised correctly, and with the wrong dilution may even turn out to be harmful.

The ultimate effectiveness of a programme in the field does not depend solely on the medical efficacy of a given intervention. The accuracy in diagnosis, compliance with treatment protocols by both patients and providers, the acceptability and accessibility of the intervention are all critical. These factors may be further analysed critically as follows:

Efficacy. It is the extent to which a specific intervention does bring about cure in patients who are diagnosed correctly and cared for in the ideal circumstances. Efficacy defines the upper limit for maximum benefit. But both behavioural and non-behavioural factors can dilute the benefits.

Diagnostic accuracy. This may be defined as detection of patients with a health problem or condition. The extent to which patients with the condition are correctly distinguished from those without the condition has a bearing on the success or otherwise of the intervention.

Provider compliance. It is the extent to which appropriate diagnosis and treatment (preventive, therapeutic or rehabilitatory) are complied with by the health worker.

Patient compliance. It is a measure of the extent to which patients comply with the health provider's recommendations and treatment.

Coverage. This defines the extent to which a service is being correctly utilised by all patients who could benefit from it. It raises the issues of acceptability and accessibility of services to patients with health needs.

When all these issues have been taken into consideration,

$$\text{Cost effectiveness} = \frac{\text{Community effectiveness}}{\text{Community costs}}$$

Financial reports

In addition to reports and statements concerned with Planning, Budgeting, Cost Control, and so on, there are other documents, for example, Balance Sheet, Income and Expenditure Statements, Statistical Reports, which are prepared from time to time by accountants and finance departments. It can be very valuable for a District Health Team to have some insight into the purpose served by such documents and what additional information can be derived from them, as an aid to a better understanding of the financial 'health' of a District. A useful question to have in mind is: 'If this were my money that was being spent, what information and checks would I want to have to be sure that my money was being used properly?'.

Economic recession and district health financing

During the 1990s and continuing from the beginnings in the 1980s the economies of most developing countries have suffered major recessions. This has been reflected in a fall in the proportion of national expenditure on health. In some countries the health budgets have been reduced by two-thirds to half.

Ministries of Health tend to respond to budgetary cuts in a predictable manner. First, the maintenance of buildings is postponed. Next to go is the repair or replacement of vehicles and equipment. Then bills get settled late. Finally, drugs and petrol suffer cutbacks. Similar cutbacks often happen in staff salaries indirectly, with increments being withheld, or staff salaries not being allowed to rise in parallel with inflation.

Countries around the world are exploring new sources of revenue. The following list is an example of different approaches:

(1) In some countries employers are being charged the full cost of treatment for accidents at work. All workers in industry, commerce, and plantations fall into this category. This approach can be further developed into a pre-paid insurance scheme to cover employees.
(2) Treatment for road accidents may be fully charged to the owner of the vehicle (and the insurance company, if insured).
(3) Charges for private accommodation in the hospital may be raised to cover the 'full' cost of treatment.

(4) Out-patient consultant clinics can have paying sessions where individual patients are seen by appointment.
(5) Patients who by-pass the local clinic and present at the hospital without a letter of referral may be charged full costs.

BUILDINGS

Buildings in the form of dispensaries, Health Centres, clinic and hospitals are part of the normal set-up for health care. There are, however, advantages *and* disadvantages in relying on buildings as a principal means of providing health care (see table 5.26).

Managers need to think very carefully before deciding to spend money on new buildings or on the extension of existing ones. A first consideration should always be: How can a health plan be developed which makes the minimum use of buildings? A survey should also be made of buildings and facilities already in the community in which health services could be provided or developed, for example,

- schools – development of child health services
  – use during or outside normal school hours
- ·missions, religious premises – may have room or under-used capacity which could be used for clinics, health education activities
- houses, homes – of village health workers, local healers or others can be adapted or extended for health purposes
- village meeting places, community development, agriculture projects – may all have useful facilities
- commercial premises – private pharmacies and drug stores
- workplaces – employers may be encouraged to provide facilities for their employees, families and immediate neighbours

By encouraging local people to extend and make use of such facilities, the community can be encouraged to think of health provision as its own responsibility, and it is also a means of strengthening links between health and other aspects of the community's life.

An assessment should also be made of existing health buildings. A worrying feature of health provision in a number of countries is the extent to which buildings put up in the past are not being fully used. Health Centres which were intended to provide a basic range of services may be under-utilised, whilst people travel long distances to overcrowded hospitals to obtain these same services, at much higher cost and inconvenience to themselves and the hospital. Such by-pass phenomena are a challenge to Health Managers and the reasons need to be investigated and put right. Often it has been the case that Health Centres were established but were not adequately staffed or supported;

Table 5.26 Buildings for health care

Advantages	Disadvantages
Provide *privacy* for personal consultations	*Expensive* both to build and maintain
Provide *security* for equipment, stores, drugs, etc.	*Distant* from many homes in the community
Provide *shelter* from rain, heat etc.	Tend to serve only the *immediate locality*
Enable facilities and resources to be *concentrated* in one place	*Isolate* staff from the community they are meant to serve
Are a *visible focus* for health care	Become an *end in themselves* rather than a means to better health
Are part of the *community* and attract local support	By their existence *separate health* from other aspects of life
Can be *silent teachers* of a clean, hygienic way of life	Encourage institutional thinking rather than problem-solving of the community's needs
Maintain *continuity* of health services as people come and go	*Inflexible* – built for one purpose and difficult to change for another
A base for *teamwork*	Can cause *problems* between staff over use of rooms, etc.
Necessary for complicated medical treatments and procedures	Encourage the use of expensive treatments and procedures

the services they provided therefore were not adequate to begin with and may have deteriorated further; the population lost confidence and no longer went to the Health Centre but went instead to the already crowded hospital. In this situation what is needed are not new buildings, but attention to the management deficiencies at the root of the problem. An assessment of existing buildings should therefore cover the following points:

Usage	are health buildings under-used or over-used?
Activities	what clinics, out-patient, in-patient, teaching activities take place? Should such activities continue? Are they scheduled and organised efficiently?
Size	is the building too big, too small? What can be done about it? Develop new activities? Extend the building?
Location	are buildings most conveniently located for access by the population? Where should any new buildings be placed?
Condition	are repairs needed? Is maintenance being regularly carried out?
Staffing	are there the right numbers and quality of staff? Is suitable training available?
Support	is there sufficient support from the local community and the health service to carry out the function expected?

Supplies	are the supplies of drugs and equipment regular, with adequate stocks of essential drugs?
Facilities/Services	does the building have the necessary furniture, equipment and services (for example, water supply) to do its job effectively?
Security	is the building safe from misuse, theft, and so on?
Flexibility	how easy is it to adapt the building to new purposes?
Operational policies	are there clear statements of how the various activities which take place within the building are to be organised?

The work involved in actually planning, designing and constructing a new health building is a study in its own right and most countries have standard plans for Health Centres, sub-centres and hospitals. A practical guide entitled 'A Model Health Centre', published in 1975 by the Conference of Medical Missionary Societies in Great Britain and Ireland, links the practical and technical aspects of Health Centre building including such aspects as Staffing, Workloads, Social Areas, Room Layouts, Storage Facilities and Record Keeping, and so on.

SUPPLIES AND STORES

An effective supplies system is essential for the smooth running of rural health care. The range of goods and equipment needed is not in itself large, but the confidence that the right goods will be in the right place at the right time is crucial. A Primary Health Care supplies system should be relatively simple and straightforward, and should relate to other supplies systems in the District, for example, public transport; systems based on a District Hospital; or other agencies like Agriculture, Community Development and so on. Greater efficiency and savings can be made by co-ordinating transport and distribution systems in remote and widespread rural areas.

Essentially, a supplies system deals with requisitioning or ordering, purchasing, receipt, care and custody, and finally, issue of goods to users. In most countries the Central Stores Department, advised and guided by a Pharmaceutical Supplies Committee at the Ministry of Health, undertakes these functions and provides a service for the Primary Health Care Programme. The Central Medical Stores at the National level has its branches and counterparts at the Regional and District level. Besides these central supplies, there may be a small allocation for local purchase of drugs not provided by the Central Medical Stores.

A centralised store at District level needs a full-time storekeeper with expertise in supervising the various stages of supplies management. Each village health worker should be issued with a box containing essential drugs, dressings, and so on. Members of the District Health Team must be familiar with the way in which the system operates to ensure that the essential requirements for providing Primary Health Care are in regular supply. Too little held in stock can lead to frequent shortages; large stocks are costly and provide greater opportunity for pilferage, deterioration, over-ordering and other abuses. In determining how much stock to hold the following factors should be taken into account:

(1) Monthly, quarterly or annual requirements for each item
(2) The price of items
(3) The time taken between placing an order and receiving the goods
(4) The purchasing cost for each order

Stores *records* are important. The *inventory* is a list of furniture and equipment for which a manager is responsible in each health building or mobile team. It should be kept up to date and checked regularly to keep control over losses, breakages, stealing, and so on. A *requisition* is an order form for obtaining goods from Central Stores; a good principle of delegation is that the person who uses supplies should have the authority to requisition them subject to supervision by the manager. Within a Central Store, records of goods held in stock will be kept in a *store ledger*, and that should agree with the number of items on the shelves of the store which are also noted on a *tally card* or bin card. As goods are received into the store, entries are made in the ledger and on the tally cards, and as goods are issued against a requisition appropriate entries are also made on a *stores issue voucher* and in a *stores issues book*.

Each Health Centre should have its strong-room for storage. Clear and well-understood supplies procedures are needed to ensure that goods are ordered and obtained before stocks run down, and a system of periodic stock-taking and auditing to ensure that goods are not lost, stolen, or misappropriated, or deteriorate through poor storage.

Where certain goods are used frequently, for example, dressings, needles, and so on, re-ordering is simplified by having a 'topping-up' system whereby a top limit of, say, 50 or 100 items is held at any one time, and stocks are replenished to this level each time deliveries are made. Control over the use of those items is obtained by setting an upper limit to the number which can be used in any given period of time.

Problems in supplies organisation need to be identified and remedied. *National* shortages cannot be dealt with locally, so what few supplies are available must be used for priority cases, and where possible alternative or locally available supplies used. Where the problem is of *transport* or *distribution* those systems need to be looked at and improvements made. Often shortages at the peripheral level are due to a poor transport and distribution system or inefficient management rather than due to national shortages. Usually problems arise through *inefficiencies* and *over-bureaucratic controls* which need to be dealt with thoroughly. *Over-stocking* and *fraud* need to be tackled immediately. If there is insufficient money available to buy necessary goods and supplies, then *budget allocations* may need to be reviewed. Managers need to develop health care and prevention activities which rely as little as possible on imported expensive supplies and equipment, and to become skilled in using locally available resources and technology for health care purposes.

DRUGS, VACCINES AND OTHER PHARMACEUTICALS

The control and eradication of many diseases in developing countries depends on the systematic and proper use of drugs and vaccines. The cost of drugs can be high, and health managers face pressures both from local populations who see drugs as the solution to their health problems, and from drug companies who wish to extend their sales. In developing countries drugs can contribute to more than 50 per cent of medical costs. At the same time, in many places great savings could be made through changes in prescribing practices. Effective systems are therefore needed to deal with the standardisation, procurement, distribution, use and control of drugs.

Management of any supply system for drugs and pharmaceuticals may be usefully discussed under three main headings: (i) Procurement, (ii) Indenting and (iii) Standardisation, storage and distribution.

Procurement is best done centrally for countries which depend on outside sources of supply. Decentralised procurement is recommended for drugs with a short shelf life or those required for the treatment of relatively uncommon diseases. It is useful if these drugs can be made available at local drug stores and pharmacies. In many cases, drugs will have to be procured from outside the District and supplied through a state-sponsored procurement agency. To ensure that high quality drugs are ordered at reasonable cost, such central procurement agencies must ensure that drugs are obtained from reputable pharmaceutical companies and at bulk-purchase prices under contract agreements.

Indenting for drugs should as far as possible be based on epidemiological information obtained from the records of local health care institutions. Drugs form an important element in the District Health Plan which shows the basic pattern of disease in the District and what measures are planned to deal with it. The drug budget is likely to be best spent on a limited range of inexpensive items from a national formulary or standard list for those health needs which are most widespread, for example, antimalarials, vermifuges, iron and vitamins, antibiotics, anti-tuberculosis and anti-leprosy drugs. The anticipated monthly usage should be worked out for each Health Centre, clinic or health worker, and supply and distribution systems co-ordinated to ensure that the right drugs get to the right place at the right time. Adjustments can be made to the quantities of drugs supplied on a month by month basis in relation to the number of cases dealt with. Drugs can be classified into *fast-moving* and *slow-moving items*. Indents for fast-moving items like antibiotics and antimalarials can be prepared in such a way that adequate monthly stock levels are maintained in local Health Centres, and hospitals as well as at District and Regional drug depots to allow free and uninterrupted flow of supplies from the National to the local level. Where possible a 'topping-up' system may be used.

Standardisation is an essential element in efficient drug management, especially at the periphery of a National Health Care system. The principle of standardisation applies to the design of simple drug formularies for use at local levels of the District health care system, depending on the level of training of staff and on the health problems they face. The use of a simple drug formulary can act as a means of checking over-prescribing and limiting the range of drugs which can be prescribed. This will reduce costs and simplify treatment procedures besides simplifying many problems of centralised buying.

Pre-packing of drugs for Health Centres, health posts, polyclinics and community clinics is a means of simplifying prescribing procedures and also acts as a control measure to check over-prescribing and pilfering.

Vaccines

Vaccines present a special problem within the District Health Service distribution system because they have to be kept at cold temperatures in order to maintain potency. The process of distributing vaccines at a constant low temperature from the point of manufacture to the point of use is called the 'cold chain'. It is a complex process as vaccines are frequently made in one country and used in another. Transport to outlying places must be carefully planned. Refrigerators need proper maintenance and reliable fuel supplies. Health Managers need to be knowledgeable about the cold chain, as detailed in the WHO Expanded Programme on Immunisation, and to review regularly the way the cold chain operates in the District.

TRANSPORT

Transport is an essential part of the communications system of a Health District.

Transport is necessary for:

Supervision and support
Distribution of supplies and drugs
Mobile teams

Public health nurses, sanitarians, health educators, family planning workers, and so on, carrying out their work in the community
Training – to take 'teachers' to 'learners' and vice versa
Patients – when referred to Health Centre or hospital for treatment
Data/Information – from villages, Health Centre, to District HQ
Salaries/Wages – for locally based staff
Visits of District Health Team members throughout the District, to other agencies, regional and provincial centres, and so on.

These needs can be considerable, and with the high costs of vehicles, maintenance and fuel, can consume a large part of a District's health budget. One way out of the difficulty is by organising services in such a way that people at local level do as much as possible for themselves, and do not travel unnecessarily outside their locality. Village health workers, by definition, live within their villages and do not need transport. Health Centres should be based in or near to larger settlements with as good road and path

Table 5.27 **Transport problems and how to tackle them**

Problem	Ways of tackling the problem
Breakdown of vehicles	Ensure regular maintenance with log book for each vehicle, showing mileage and type of maintenance required.
	Develop good relationships with local workshops, garages, mechanics.
	Plan for temporary substitution of vehicles, bicycles, and so on.
	Train users in proper upkeep and use of their vehicles.
	Keep a reasonable supply of spares.
	Regular vehicle replacement.
Misuse of vehicles	Good supervision and training.
	Good scheduling of vehicle use, so fewer opportunities for misuse.
	Restrict use to named individuals.
	Effective controls, log-books.
	Enforceable policies on private use.
High transport costs	Examine transport needs – are all journeys really necessary?
	Use low cost transport – bicycles, foot. (A health worker, including doctors and nurses, on a bicycle or on foot gets closer to the community and becomes better known.)
	Standardise vehicles, bicycles and spares.
	Re-examine systems of work in the District.
	Use public transport.
	Work out cost per mile, and transport costs for different services.
	Share transport with other agencies.

communications as possible. Many members of the outreach team at Health Centres, for example community nurses, midwives and so on, may be encouraged to use bicycles. Communication and transport needs in the District should be properly co-ordinated. Thus a weekly or fortnightly return journey by a multi-purpose vehicle between the District HQ and each Health Centre could be used to combine supervision and support, distribution of supplies and drugs, visits of public health nurses and others for specialist clinics, payment of wages, returns of data and information, and for training.

Appropriate resources need to be put into transport. A well-maintained 4-wheel-drive vehicle may be essential to a well-run District, so that contact can always be maintained throughout the community, but other transport needs may be met by simpler means, for example, bicycles, motor cycles, foot or boat and public buses. The principle applying to transport is that simple health care requires simple, not sophisticated support.

As the entire work of the District can break down if the transport system breaks down, likely problems should be identified and ways of dealing with them worked out (see table 5.27)

FURTHER READING

Cockman P., Evans B., Reynolds P. *Client – centred Consulting. A Practical Guide for Internal Advisers and Trainers.* McGraw-Hill, London, 1992.

Drucker P. F. *Managing for Results.* Heineman, London, 1989.

Johnstone P., Ranken J. *Managing Health Centres for Primary Health Care.* FSG Communications, Cambridge, 1994.

McGrath S. J. *Basic Managerial Skills for All.* Prentice Hall of India Ltd. New Delhi, 1980.

Pearson A. *Medical Administration for Front-line Doctors.* FSG Communications, Cambridge, 1990.

World Health Organization. *Managerial Process for National Health Development: Guiding Principles.* WHO, Geneva, 1981.

6 Getting Feedback: Monitoring and Evaluation

Evaluation is a systematic way of learning from experience so as to improve current activities and promote further learning. In the case of health programmes, the objective of evaluation is to improve the services for delivering health care and to guide the allocation of resources. Thus evaluation is closely linked with decision-making both at the operational as well as at policy level.

Evaluation asks five basic questions:

(1) What was intended to happen (objectives)?
(2) What has actually been achieved so far compared with objectives?
(3) In the light of such a comparison what value should be placed on the methods used (process)?
(4) What use can be made of the information gained from questions 1, 2 and 3 (feedback)?
(5) What is the whole exercise teaching us about managing future activities?

Evaluation is a learning process. The valuing, feedback and learning functions of evaluation offer the means of checking, correcting and improving the activities of the District Health Organisation. As such they comprise an important facet of the work of the District Health Management.

From the practical angle the same questions asked about community problems and resources as described in chapter 2 can also be asked about evaluation. What should be evaluated? How should it be done? By whom? Where? and When?

WHY DO WE NEED FEEDBACK?

Feedback is needed to find out what's going on, to avoid difficulties and problems, and to have information at hand for decisions as they crop up.

Getting feedback (or evaluation) is one way of finding out if we are on the right path, and whether health programmes and activities are meeting the needs for which they were drawn up. There is a Ghanaian proverb which says 'If whilst clearing the bush to make a path one does not look back to see where the path is heading in relation to the starting point, one may end up with a path which goes completely round in a circle'. This is why we need feedback about whatever we are doing. This on-going self-evaluation is the type of feedback most frequently used in management. It is usually small in scale and short in time. What needs to be monitored is frequently changing depending upon the several activities going on. Information needs to be gathered continually to give an indication of what is happening at the time (see table 6.1).

There are three other forms of evaluation of health care which are frequently performed and which it is helpful for District Health Teams to know about, even though they may not routinely be involved in them. One type assesses a situation before a project begins. It takes a measure of the existing health

Table 6.1 **Why do we need feedback?**

Type of feedback	Reasons for its use
Assessment of current situation (formative evaluation)	Prior to a project 'Needs assessment' Evaluation of current situation
Assessment of plans (project appraisal)	Are project plans appropriate? – to needs? – to country? – to the people being involved? Political mileage? Blockages for economic reasons?
Intermittent or end of programme (or a particular point in time) (summative evaluation)	Has the project met its long-term objectives? Has the general policy been implemented? Has the donor agency policy been implemented? Shall we continue or discontinue the programme? Shall we replicate the programme elsewhere?
On-going self-evaluation (situational evaluation)	Small-scale, short time period Used by managers and by participating personnel and by communities Must be continually changing and geared directly to short-term action Can improve strategies and techniques

services, their utilisation and the health status of the community. It is sometimes called 'needs assessment' or 'formative' evaluation. A second type assesses plans for a project. It is sometimes called 'project appraisal' or 'pre-implementation appraisal'. Programme organisers assess whether a project is appropriate to the identified needs, appropriate to a country and appropriate to the people being involved. This type of evaluation is sometimes vulnerable to people trying to gain political advantage by arguing for a particular type of plan although the country cannot afford it. Alternatively, planners may assume that money is all that matters and may forget they need people who are committed to a plan if it is to be put into action. A third type of evaluation occurs at the end or at a defined stage of a programme or possibly intermittently while the programme is running. It is sometimes called 'summative' evaluation. Its aim is to assess whether a programme has met its long-term objectives, both at a general policy level as well as in detailed outcome measures. It is based on data about outcomes such as morbidity and mortality rates which may only change over long periods of time. Donor agencies often particularly ask for this type of evaluation because it can give a clear idea of what specific gains have been made, and whether their general policy has been put into effect. Finally, all activities and programmes need on-going evaluation by their performers to improve strategies and techniques. All of us concerned with improving the effectiveness of our work are continually doing this but mostly unconsciously or informally. In order to integrate evaluation into the day-to-day work of the manager of the Health Team, it has to be planned for and included in the job description of the manager.

WHAT IS MONITORING?

Monitoring is different from evaluation. Whereas evaluation takes place at agreed intervals (mid-term or end of term evaluation) monitoring is performed on a day-to-day and on-going basis. Evaluation is asking 'Have we arrived at the defined objective?'. Monitoring is asking 'How well are we progressing?'.

The District Health Manager needs to know if all the activities scheduled are actually taking place, if standards of work are being maintained, and if there are problems or difficulties which need attending to. In the District, activities are going on in many places, and the Manager has to rely on other people for the execution of activities. The Manager cannot be in every possible place to know what is going on. If the District is well organised, and there has been rational delegation of work, the routine demands will be well met by departmental heads and supervisors. But what the Manager needs to know are the more serious and deep-rooted problems at different service points. He needs to have a more comprehensive, and broad view. To have such a view accurate and up-to-date information of what is going on in the

District is needed. Thus monitoring is gathering information to know whether the previously agreed course is being maintained, resources are wisely spent and staff are utilised effectively. Minor and continuous changes to a programme for keeping it on course are more effective than major changes when the situation is really serious. Besides keeping the District Manager's knowledge about the activities in the District up-to-date, monitoring has the added advantage that new opportunities can be exploited as they arise, and problems may be identified early before they get out of hand. Monitoring is different from policing! The purpose is entirely supportive of one's staff and of a previously agreed plan. If carried out in the right spirit monitoring encourages staff to give of their best, because they know that their supervisors are taking interest in their work, recognise their achievements and are at hand to deal with their problems.

The following principles of effective monitoring will help to make the process more structured and standardised:

- Determine what to monitor, and at what interval.
- Determine how to monitor.
- Develop a checklist.
- Monitor as planned.
- State any problem identified.
- Describe each problem.
- Identify possible causes of each problem.
- Identify and implement the solutions.
- Monitor the solutions.
- Provide feedback to the health worker.

The problems of shortages because of delayed ordering, absenteeism, poor maintenance of equipment, lack of cleanliness or orderliness at service points, missed opportunities for containment of epidemics or environmental hazards can be largely avoided through regular monitoring and control.

Methods of monitoring

Reports   Written or verbal concerning particular aspects of work, any special problems or difficulties. Written reports should be kept brief with specified headings. The manager or supervisor must give immediate feedback to the person reporting.

Visits   Visits are particularly important, because they add the human touch besides the first-hand knowledge obtained. They help to boost the morale of outlying communities and health staff.

Checklists   A checklist ensures that all the relevant aspects of work have been enquired into, and there has been no oversight because of other pressing problems.

Meetings   Meetings are important to review and alter local targets, for discussing reports and for future planning.
Complaints   In all human interactions there are bound to be some disaffected people. In general complaints are good indicators that there is something wrong. They must always be followed up. Complaints are one way of identifying problems at an early stage.

Monitoring and evaluation is one way of finding out why health plans are not working as well as hoped for, and that modifications are needed. Environmental demands, and consumer choice are as unpredictable as resources. By responding to the environmental factors and learning from experience, an adaptive approach in management can be promoted.

The frequency with which monitoring is done will obviously vary with the level of management. The nursing sister in charge of a ward or theatre monitors her junior nurses on a day-to-day basis, and so does the consultant for his house officers. The matron of a hospital may monitor the work of her staff at monthly intervals. The manager of the EPI programme monitors the supervisors quarterly, and the District Health Manager monitors the health activities of the District biannually. A regular system of monitoring gives a framework to the health activities in the District, and maintains them on course. It also provides the regular pulse for all the health work. But that is not all. Good managers are constantly alert for the signals in their daily round of contact that tell them of irregularities.

From time to time opportunities may arise for detailed monitoring because of occurrences like outbreaks of food poisoning (monitoring of food hygiene and water supply), of infections (monitoring of EPI or communicable disease control), of disciplinary problems. Or there may be annual events like preparing the next year's plan, or the budget or poster displays for visitors.

Inadequate monitoring and control happens if individuals are assigned the responsibility without knowing the basics of management, or because of long distances with non-availability of transport. A proper information base may not have been created for the District, or the systems of monitoring may be archaic (for example from colonial times) and cumbersome.

WHAT SHOULD BE EVALUATED?

The next most important question needing an answer is *what* should be evaluated?

What needs to be evaluated depends on the community diagnosis of problems and resources
Community diagnosis (chapter 2) identifies certain problems and potential

resources. Evaluation will be able to answer questions such as 'Have the disease problems been altered?' 'What are the new disease problems?' 'Are resources being used effectively?' 'What new resources might be mobilised?' 'Have the underlying social causes and other determinants of ill health been fully identified?' 'Are the health services well geared to tackle the health problems experienced by the majority of the people in the area?'.

What needs to be evaluated depends on the plan of action for the health team in the district

The health plan (chapter 3) identified specific goals and targets as well as strategies for action. Evaluation of the plan's effectiveness is to answer questions like: 'Have the plan's objectives been fully realised?'; 'What tasks need to be done better?'; 'How efficiently are things being done?', in other words the cost-effectiveness of the interventions employed; 'How effective have the plan activities been?', in other words 'Are the desired outputs being obtained and if so, are the outputs useful?' and therefore, 'Is the programme needed?'. These questions point to the six levels of evaluation (see table 6.2). Thus, during evaluation one examines the current processes in use (process evaluation) as well as the outcomes and consequences of the activities undertaken (outcome evaluation).

In health care evaluation one is continually weighing the process against the outcome. Process depends upon the settings where care takes place including buildings, equipment, skills of workers, administrative support, and so on. Structure, process and outcome form a triangular relationship, and have to be taken into account for evaluation (see table 6.3). An underlying assumption linking all three is that the particular medical technology is efficacious.

What are the key components of evaluation which often get neglected?

It has recently been recognised that there are a number of key components of the effective provision of a District's Health Service which are frequently neglected during evaluation of District Primary Care. These are accessibility and coverage, community participation and community development, functional integration and support within the health service, feasibility in terms of

Table 6.2 **Six levels of evaluation**

Level 1	Activity: Is the planned activity working?
Level 2	Quality: Is it meeting the quality standards?
Level 3	Efficiency: Can the cost per unit output be improved?
Level 4	Effectiveness: Is it producing the outputs desired?
Level 5	Outcome validity: Are these outputs useful?
Level 6	Overall desirability: Is the programme needed?

Table 6.3 **Structure, process and outcome**

Structure
The quality of health care is assessed through a study of the settings in which the care takes place. This includes adequacy of facilities and equipment, administrative processes, organisation, qualifications and skills of the medical and nursing staff. The assumption is that given proper settings, good medical care will follow.

Process
This considers the standard of care: clinical history; physical examination; diagnostic tests; scientific basis for diagnosis and therapy; co-ordination and continuity of care; patient and provider compliance. The assumption is that given the proper procedures, good health outcome will result.

Outcome
Outcome considers whether a change in a person's health status is attributable to health care received. It examines recovery, restoration of function and survival. Often there are multiple factors which affect health outcome besides the treatment received.

---

cost, and quality of services. These components are crucial to the success of a health programme. Ways and means of measuring these components need to be included in all evaluations of District health activities.

Community participation in evaluation is an integral part of community involvement in health (page 102). True participation happens when following free discussion and decision-making the community takes initiative in taking action, and has a full say in the ensuing interventions as well as their evaluation. Such full participation is rarely achieved because each community is unique. But it is something to aim for. Once a relationship of trust is established and a pattern of constructive criticism develops, evaluation can only strengthen the health programme. The distinguishing features of participatory evaluation are appreciation by the people of their contribution to the success of health programmes, the improvement of their knowledge on health matters, the development of skills in solving problems affecting their lives and health, and thereby a growing commitment to the health activities. Participatory evaluation is for creating a sense of ownership of local health activities. Its merits lie in the improved understanding and boosting of morale for the health workers and community alike. Balanced against this is the fact that the results may be less satisfactory in terms of statistical analysis. However this should not be a major difficulty if adequate baseline data are recorded at the outset with quantifiable objectives.

A team of elected representatives may be used to steer the evaluation. The team may need help to focus on important problem areas, and to identify possible solutions. Perceptions of problems are often very different between the providers and the beneficiaries. Success can be demonstrated in quantitative terms only when the objectives are quantifiable. But things like attitudes, motivations, leadership qualities, goal priorities and communication cannot be evaluated precisely even though they are so important for success. Leadership skills are needed to control contentious issues which can come up and waste time.

In participatory evaluation the findings are usually understandable by the community and immediately usable. Improvements in technology like laptop computers mean that all data analysis can be accomplished at the community level.

WHAT ARE THE ESSENTIAL ELEMENTS OF PRIMARY HEALTH CARE WHICH NEED TO BE EVALUATED?

There are eight essential elements of Primary Health Care (PHC) which will obviously need to be evaluated (see figure 6.1). These are: food supply and nutrition, water and sanitation, mother and child health, immunisation, prevention and control of locally endemic diseases, management of common illnesses and injuries, provision of essential drugs, and community mobilisation and awareness. The way of evaluating PHC has now been made much easier because there is both a global analysis of what these essential elements entail and several country examples specifying precisely what the tasks under each of the eight elements are. Evaluation is then simply a question of identifying these tasks and then seeing to what extent these tasks are being put into practice (see tables 6.4 and 6.5).

Indicators for monitoring 'progress towards health for all' are being discussed at an international level and include indicators of health policy, socio-economic indicators, indicators of health status and indicators of health care provision (WHO, 1980). Health status indicators suggested include proportion of infants born with low birth weight, height and weight of children, proportion of pre-school children with small arm circumference, infant mortality rate, child mortality rate, under-five mortality rate, under-five proportionate mortality, life expectancy, maternal mortality rate, crude birth rate, disease-specific death rates, proportionate mortality from specific diseases, morbidity incidence and prevalence rates, and prevalence of long-term disability. Suggested indicators of provision of health care include coverage, physical accessibility, percentage of population served, socio-economic accessibility and population ratio to health personnel. In addition there has been a growing realisation of the need to find out what is happening at the 'front-line' of the health care system and to devise appropriate conceptual and analytical methods for doing this (WHO, 1981). This requires

264   *District Health Care*

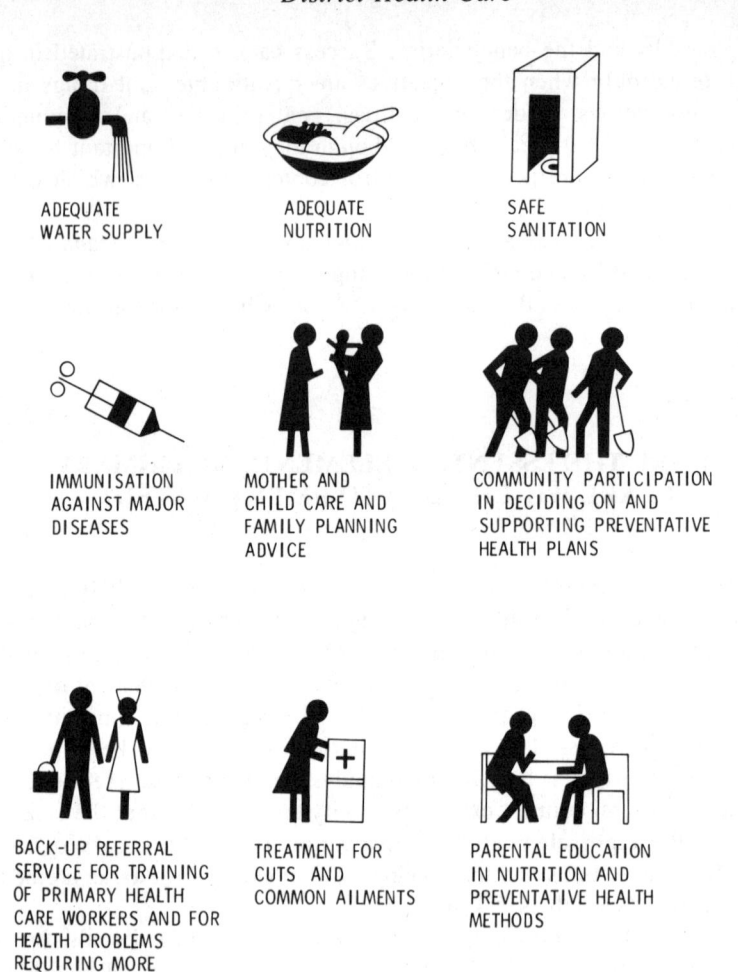

ADEQUATE WATER SUPPLY	ADEQUATE NUTRITION	SAFE SANITATION
IMMUNISATION AGAINST MAJOR DISEASES	MOTHER AND CHILD CARE AND FAMILY PLANNING ADVICE	COMMUNITY PARTICIPATION IN DECIDING ON AND SUPPORTING PREVENTATIVE HEALTH PLANS
BACK-UP REFERRAL SERVICE FOR TRAINING OF PRIMARY HEALTH CARE WORKERS AND FOR HEALTH PROBLEMS REQUIRING MORE QUALIFIED CARE	TREATMENT FOR CUTS AND COMMON AILMENTS	PARENTAL EDUCATION IN NUTRITION AND PREVENTATIVE HEALTH METHODS

Figure 6.1   The essential elements of primary care

the orientation of front-line workers to think of disease, disability, accidents as well as vital events like pregnancy, births, deaths and so on, in terms of the total population in which they occur, the area and time in which they happen and the group of people affected. The survey method has been the mainstay of the methodology of evaluation, but it has recently come under criticism. Because of the difficulties of travel to the more remote hamlets and villages of the District, especially during the wet season, many of the rural problems go unseen. Often, survey teams do not stay overnight in the area; much of the working day is spent in travel, and during the brief period spent in the village much of their time is taken up by local reception committees. Individuals and

Table 6.4 **Evaluation – what it should, can and can't do**

What it should do	What it can do	What it can't do
Should come out of data collected	Use specified criteria	Can't all be done objectively. Much must be subjective
Evaluate not the project but the process of providing care		Decision-making is also based on some assessment of what is likely to work or not, usually influenced by powerful people which in turn affects results and outcome
	Indicate where change is desirable (can bring change)	Does not necessarily effect change about every item
		Cannot necessarily persuade staff to support evaluation methods and results, especially if
		– it is 'external' evaluation
	Give sense of direction and commitment (if done corporately)	– workers think it is 'snooping'
		– it appears to be threatening: people need to think some good is coming out of evaluation, otherwise will cause disappointment or raise expectations.
	Can raise morale	
	Good for training	Need to have initial hunches about what directions are possible

Table 6.5 **Uses of information: good and bad**

+	−
It can be of direct relevance to the work	Remove the fear that evaluation results will be used by people in authority 'up there' to criticise their work
If backed up by people with authority it helps to get things done	
Can indicate whether anything is being done about problems identified	
Can be made part of the process and not a one-off exercise	

families with problems get pushed into the background and unless a special effort is made to identify them, they are likely to go unnoticed and their problems unrecorded.

WHICH LEVEL? WHICH COMPONENT? – LEVELS OF EVALUATION (see figure 6.2)

In a District Health Programme several components, tasks and activities of health facilities may need evaluation. On the other hand, the evaluation may be at the level of a village, a group of villages or the entire District. Thus one may conceive of a hierarchy of evaluation, as follows:

- District, health station or local community
- Programme, project or local service
- Components of a programme, for example, MCH, family planning, communicable disease control *or*
- Comprehensive, for example, Primary Care or District Health Service
- Health status, health care or determinants of ill health
- Care of high risk groups

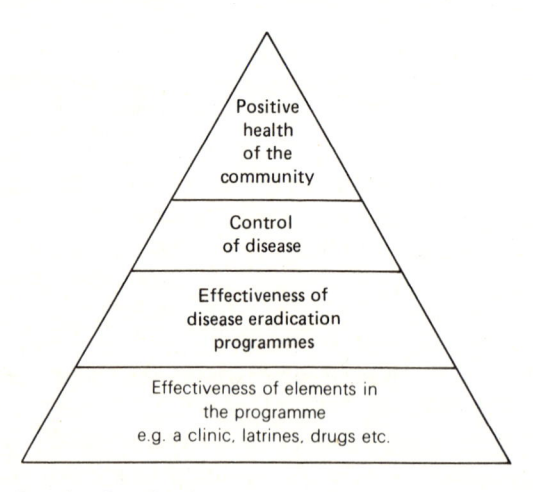

Figure 6.2 Levels of evaluation

(Pyramid, top to bottom: Positive health of the community / Control of disease / Effectiveness of disease eradication programmes / Effectiveness of elements in the programme e.g. a clinic, latrines, drugs etc.)

WHAT LEVEL OF FEEDBACK AND WHICH COMPONENT IN A PROGRAMME?

Feedback can take place at a number of *different levels*; District, health station and local community to name but a few. It may focus on a programme, a project or a local service. A *single programme* may be evaluated such as mother and child care, communicable disease control, family planning, or environmental health. Alternatively there may be *comprehensive* evaluation such as of

Getting feedback 267

primary care or a District's over-all activities. As outlined in chapter 2, evaluation may identify the health status, the health care provision or the causes of ill health.

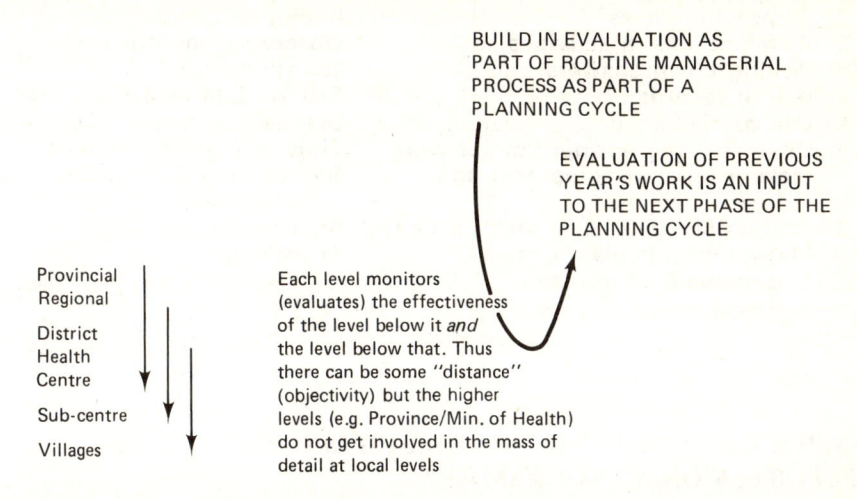

```
                                    BUILD IN EVALUATION AS
                                    PART OF ROUTINE MANAGERIAL
                                    PROCESS AS PART OF A
                                    PLANNING CYCLE

                                         EVALUATION OF PREVIOUS
                                         YEAR'S WORK IS AN INPUT
                                         TO THE NEXT PHASE OF THE
                                         PLANNING CYCLE

Provincial         Each level monitors
Regional           (evaluates) the effectiveness
District           of the level below it and
Health             the level below that. Thus
Centre             there can be some "distance"
                   (objectivity) but the higher
Sub-centre         levels (e.g. Province/Min. of Health)
                   do not get involved in the mass of
Villages           detail at local levels
```

Figure 6.3   Levels of feedback and which component in a programme

WHICH SPECIFIC QUESTIONS CAN FEEDBACK FROM A HEALTH CARE PROGRAMME CONSIDER?

Feedback can consider one or two major topics out of a number of important issues. Such topics include feedback about the appropriateness and relevance of a programme, on the adequacy of services provided (this is done by comparing services with the needs for them), on effectiveness, on efficiency and on impact. Another important topic for feedback is the flexibility of a programme to respond to new directions, initiatives and needs identified while it is in progress. Acceptability of new approaches or directions can be measured with regard to the Ministry, the staff and the local people. Side effects and safety always need to be considered as well as the ethics of any feedback procedure.

HOW CAN DATA BE OBTAINED TO FIND OUT WHAT IS GOING ON?

Some of the possible sources of data have been considered in chapter 2. Other commonly listed procedures are listed in table 6.6.

Table 6.6 **Commonly listed procedures for obtaining data**

Procedure	Formal terminology
1 Asking, talking, recording	Interviews
2 Comparing views	Rating
3 Watching health activities	Observing unobtrusively
4 Writing down answers	Questionnaires
5 Use of records	Record data analysis
6 Use of diaries	Diaries
7 Observing the normal flow of work	Time and motion studies
8 Making a list of equipment and drugs consumed	Stock-taking and making inventories
9 Discussions with community groups	Rapid rural appraisal
10 Measuring (people, rates, specimens, foods etc.)	Surveying

QUICK OR LONG? AND WHAT DISCIPLINES IN GETTING FEEDBACK ON A PROGRAMME?

Feedback can range from 'a quick look and a good listen' to a long-term detailed investigation. It can attempt to be objective (using outsiders) or to be enlightening (when local insiders are essential). A number of disciplinary perspectives can be involved, economics (which is often quick), epidemiology and sociological (which often rely on surveys and statistical analysis), anthropological (which often requires a longer time and uses key informants for information) and political (which often relies on a non-participatory observer).

An example of 'a quick look and a good listen' is to be found in the 'community round' information described in chapter 2. The economic perspective is shown in the figures for health facilities, personnel and in the costing of salaries (all in chapter 3). Epidemiological methods show the pattern of high risk families and communities (see chapter 2), and anthropological data show for example how the women's food cropping has changed and the reasons for it (chapter 3). Discussion with people is often necessary to supplement formal surveys and to understand the reasons why certain things are happening (see tables 6.7 and 6.8).

If information collection is to be effective then those collecting it at the local level must have a first commitment to using it themselves for decision-making, in addition to forwarding it to another level to be used by others. Thus health care providers need to learn simple methods of collection and analysis of health data so that they can act on the information they receive. It should be possible for District health staff to analyse health information using simple

Getting feedback

Table 6.7 **How is feedback done?**

Quick look and listen or long term and detailed
By outsiders ('objective') or insiders ('enlightening')
Using tools of economics (usually quick, e.g. 'hard' costing of programmes and activities)
Using tools of epidemiology and sociology (e.g. patterns, surveys or identifying power groups)
Using tools of anthropology (participating observer attempting to define reasons for certain events)
Using tools of political science (non-participant observer)

Table 6.8 **What specific questions need to be answered for proper feedback?**

Questions	Comparison
Appropriateness and relevance	Comparison with policy (chapters 2 & 3)
Adequacy, quantity and cost	Comparison with needs (chapters 2 & 3)
Effectiveness	In meeting aims (chapter 3)
Efficiency	In working methods
Impact	On specific problems
Progress flexibility	In relation to new requirements
Acceptability	To local people, to staff & to Ministry
Accessibility and distribution	Of services in relation to population and tasks to be done
Side effects and safety	
Ethics of service provision and of evaluation	
Is the feedback being evaluated?	
. how much will it cost?	
. what is unsaid?	
. what is unwritten?	
. how narrow or comprehensive is it?	
. are the results discussed?	
. with whom?	
. why are those specific people doing an evaluation?	

arithmetic (and pocket calculators if available) so that it is not necessary to send information to national computer centres for analyses which may take years and which will yield results long after such information has ceased to be useful for corrective action. At present much health information gathered is put away without being processed because the people who collect the information cannot find immediate use for it.

Table 6.9 **List of epidemiological tools and techniques for frontline health workers**

A *Tools and techniques dealing with health events*
1 Listing of collective health events, such as epidemics.
2 Mapping of health events in epidemics.
3 Tally sheets for morbidity, death and attendance data.
4 Pictorial records for communication of health events by illiterate persons.
5 Cards for manual classification of health data:
   - marginal perforation cards;
   - visual selection feature cards.

B *Tools and techniques dealing with space and environment*
1 Maps to show physical environment, population settlements, safe water sources, location of health events etc.

C *Tools and techniques dealing with time*
1 Time charts and graphs:
   - of vital events;
   - of diseases and other health conditions of local importance;
   - of health activities e.g. vaccine depots.
2 Time charts to be used as reference for assessing the magnitude of current events:
   - median curves of monthly incidence of specific endemic diseases like malaria or typhoid fever;
   - graphs indicating maximum and minimum observed incidence in each month in previous years;
   - graphs of monthly activities for the current year used as a reference for judging how a pre-determined target is being reached.

D *Tools and techniques dealing with individuals and their records*
1 Road to Health cards for children on which growth data and health events are plotted.
2 Mothers' cards.
3 Family folders.
4 Checklists of community health risk factors e.g. unfavourable environment or habits, low immunisation levels, high proportion of high risk pregnancies.
5 Guides for simple surveys on health status, diet, health care coverage. These can be facilitated by sets of cards or booklets with easily recognisable figures.
6 Simple indicators of economic status (e.g. possession of a radio set), of hygienic practices and other family characteristics.
7 At-risk register of problem or vulnerable families and individuals. Special care registers.

E *Tools and techniques dealing with population and community*
1 Local census. Household surveys.
2 Practical use of population 'models' as a way of estimating the size of population sub-groups (e.g. number of pregnant or lactating women; number of infants; number of children under the age of five etc.), when only the total population figure for a community is known.
3 Village profiles.

F *Tools and techniques for dealing with health care events*
1 Tally sheets for counting attendances for health care or immunisation.
2 Cards for manual classification of health care events.
3 Calendar boxes for preparation of clinics or visits, and for tracing defaulters.
4 Drug consumption forms to help identify sudden increase in demand (for forecasting epidemics), or variation in the utilisation of services.

G *Tools and techniques of multiple epidemiologic use*
1 The community health workers' diary for the recording of unusual events for discussion with the supervisor.
2 The record books of peripheral health units.

H *Tools for supervision of health workers in the use of epidemiology*
1 Checklists of epidemiological abilities required from each cadre of health workers.

EXAMPLES OF DATA COLLECTION FORMS AND SYSTEMS USED IN LOCAL COMMUNITIES

Some local community workers may like to use illustrated data collection sheets as shown in this example from Kenya (see figure 6.4). Others may like to use the wooden compartmentalised box and colour counters used in Papua New Guinea, or use the tally sheets as in Tanzania (see figure 6.5 and table 6.9).

Records can be designed so they can be marked by non-literate people or filled in with the help of a schoolchild who can read or write.

Traditional birth attendants (TBAs) can also be taught to use coloured marbles representing well-defined and easily identifiable clinical conditions associated with pregnancy and delivery and to drop these into a container as and when such clinical conditions arise. In the Danfa Project in Ghana, traditional birth attendants have used cards of specified colours to refer patients with complications of pregnancy to midwives at Health Centres. TBAs can also record live births with a grain of maize dropped in a jar and stillbirths with a pebble put in a jar.

Rapid appraisal techniques are now being widely used. Some like the indirect method of measuring maternal and child mortality have been mentioned on page 49. Others like the 10-Questions method for assessing the prevalence of childhood disability, the use of 'key informants' for collecting information of a social and ethnographic nature, reporting from 'sentinel' health facilities as proxy measures for District level data, use of community health workers as health scouts to gather information about vital events in the community on a routine basis, conducting case-control studies on relatively small samples need to be employed more widely.

CHOICE OF EVALUATION METHOD

Experiment design.
Using experimental and control groups Experimental group gets programme, Control group does not.

Experimental Model

	Before	After
Experimental	a	b
Control	c	d

If the difference between a and b is greater than difference between c and d, then the programme is a success. The greater the difference the more successful the programme.

Quasi-experimental design

This does not require strict adherence to the principles of the experimental design.

This is a more practical way of evaluating a programme but one must control for external factors (or variables) likely to influence the evaluation.
Some examples:

(a) *Time series design*

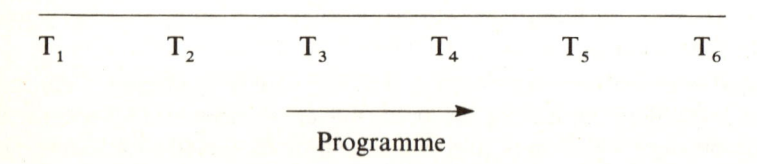

$T_1 \quad T_2 \quad T_3 \quad T_4 \quad T_5 \quad T_6$

→ Programme

Information is collected at specified periods during the course of implementation of the programme, and all changes are noted. The assumption made is that all changes and improvements are attributable to the programme. The pitfall of course is that changes may occur as a result of the mere passage of time.

(b) *Multiple time series*

This is similar to (a) above except that the observations are made at several centres where similar health programmes are being implemented.

(3) *Non-Experimental Design*

Three examples are:
(a) Before and after – study of a single programme
(b) After only
(c) After only with a control group.

Health services research (HSR)

Health Services Research (HSR) is a discipline which uses sociological, economic and other analytical methods to study health services in terms of demand, supply and utilisation of health care and how they relate to each other.

An important aspect of health services research is Operations Research (OR), which is a systematic study by experiment and observation of systems like those in a health programme. The outcome of an operations research exercise is used to improve efficiency and effectiveness of health providers with regard to acceptability, availability, accessibility and utilisation of services by the consumers.

Patient care evaluation is part of HSR. It uses three common methods to assess and evaluate the quality of care for patients. These are:

(1) *Structure* of the institution viz.
   (a) Staff
   (b) Equipment
   (c) Buildings and so on.

The limitation of this method of evaluation is that the mere presence of staff, equipment and building does not guarantee access and availability, even though the better the facilities the better the quality of care.

(2) *Process.* Here the results of interactions between patients and workers are observed either directly or reviewed by means of clinic records. This type of approach is more reliable because contact between providers of care and clients is an indicator of the type and quality of the health activity.

(3) *Outcome.* This monitors changes of the health status of the patients and the community which may be attributable to programmes carried out by health institutions serving the community. This only pays off at a high level, for example, National or Regional, and over a long time scale.

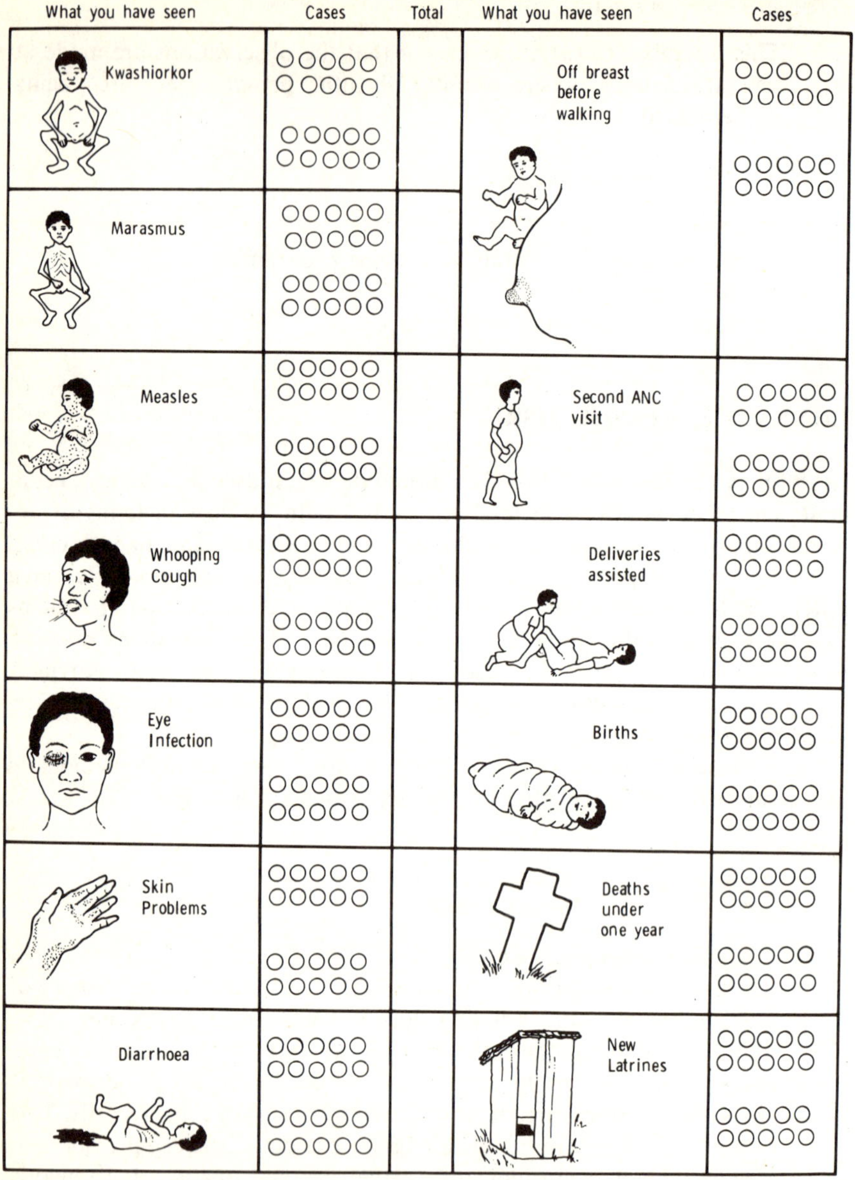

274   *District Health Care*

What you have seen	Cases	Total	What you have seen	Cases
Kwashiorkor	OOOOO OOOOO / OOOOO OOOOO		Off breast before walking	OOOOO OOOOO / OOOOO OOOOO
Marasmus	OOOOO OOOOO / OOOOO OOOOO			
Measles	OOOOO OOOOO / OOOOO OOOOO		Second ANC visit	OOOOO OOOOO / OOOOO OOOOO
Whooping Cough	OOOOO OOOOO / OOOOO OOOOO		Deliveries assisted	OOOOO OOOOO / OOOOO OOOOO
Eye Infection	OOOOO OOOOO / OOOOO OOOOO		Births	OOOOO OOOOO / OOOOO OOOOO
Skin Problems	OOOOO OOOOO / OOOOO OOOOO		Deaths under one year	OOOOO OOOOO / OOOOO OOOOO
Diarrhoea	OOOOO OOOOO / OOOOO OOOOO		New Latrines	OOOOO OOOOO / OOOOO OOOOO

Figure 6.4   Data collection sheets for use by village health workers in Kenya
Source: Adapted from Schaffer, AMREF, Kenya.

Getting feedback 275

What you have seen	Cases	Total	What you have seen	Cases
Immunizations	OOOOO OOOOO OOOOO OOOOO		Health teaching in groups	OOOOO OOOOO OOOOO OOOOO
Improved water sources	OOOOO OOOOO OOOOO OOOOO		Home Visits	OOOOO OOOOO OOOOO OOOOO
New type of food crop harvested	OOOOO OOOOO OOOOO OOOOO		Child-to-child teaching	OOOOO OOOOO OOOOO OOOOO
New tiled or metal roof on home	OOOOO OOOOO			
Child newly stopped attending primary school	OOOOO OOOOO OOOOO OOOOO OOOOO OOOOO			
Child newly started attending primary school	OOOOO OOOOO OOOOO OOOOO			

Routine health data

Collection of routine health data and their analysis is basic to the development of health services. Just as a trader has to know about his clients in order to develop his business so also does the provider of health care. Clinics, health centres and hospitals are the main venues for the collection of health data, but what is seen may not always be a true reflection of the

District Health Care

MOTHER AND CHILD HEALTH CLINIC

Name of clinic _____ Date _____

ATTENDANCE

CHILDREN

FIRST ATTENDANCE				THIRD ATTENDANCE				
00000	00000	00000		00000	00000	00000	00000	
00000	00000	00000	Total	00000	00000	00000	00000	Total
00000	00000	00000		00000	00000	00000	00000	
00000	00000	00000		00000	00000	00000	00000	
00000	00000	00000	_____	00000	00000	00000	00000	_____

SECOND ATTENDANCE				FOURTH AND SUBSEQUENT ATTENDANCE				
00000	00000	00000		00000	00000	00000	00000	
00000	00000	00000	Total	00000	00000	00000	00000	Total
00000	00000	00000		00000	00000	00000	00000	
00000	00000	00000		00000	00000	00000	00000	
00000	00000	00000	_____	00000	00000	00000	00000	_____

MOTHERS

FIRST ATTENDANCE				THIRD ATTENDANCE				
00000	00000	00000		00000	00000	00000	00000	
00000	00000	00000	Total	00000	00000	00000	00000	
00000	00000	00000		00000	00000	00000	00000	
00000	00000	00000		00000	00000	00000	00000	
00000	00000	00000	_____	00000	00000	00000	00000	_____

SECOND ATTENDANCE				FOURTH AND SUBSEQUENT ATTENDANCE				
00000	00000	00000		00000	00000	00000	00000	
00000	00000	00000	Total	00000	00000	00000	00000	Total
00000	00000	00000		00000	00000	00000	00000	
00000	00000	00000		00000	00000	00000	00000	
00000	00000	00000	_____	00000	00000	00000	00000	_____

IMMUNIZATION

CHILDREN

BCG	00000 00000	00000 00000	00000 00000	00000 00000	00000 00000	00000 00000	00000 00000	00000 00000	Total _____
SMALLPOX	00000 00000	00000 00000	00000 00000	00000 00000	00000 00000	00000 00000	00000 00000	00000 00000	_____

	1st Injection			2nd Injection			3rd Injection		
DPT	00000 00000 00000 00000 00000	00000 00000 00000 00000 00000	Total _____	00000 00000 00000 00000 00000	00000 00000 00000 00000 00000	Total _____	00000 00000 00000 00000 00000	00000 00000 00000 00000 00000	Total _____

	1st Dose			2nd Dose			3rd Dose		
POLIO	00000 00000 00000 00000 00000	00000 00000 00000 00000 00000	Total _____	00000 00000 00000 00000 00000	00000 00000 00000 00000 00000	Total _____	00000 00000 00000 00000 00000	00000 00000 00000 00000 00000	Total _____

MEASLES	00000	00000	00000	00000	00000	00000	00000	00000	_____

MOTHERS

	1st Injection			2nd Injection			3rd Injection		
TETANUS	00000 00000 00000 00000 00000	00000 00000 00000 00000 00000	Total _____	00000 00000 00000 00000 00000	00000 00000 00000 00000 00000	Total _____	00000 00000 00000 00000 00000	00000 00000 00000 00000 00000	Total _____

Getting feedback 277

		ILLNESSES DIAGNOSED							
C H I L D R E N	MALNUTRITION	00000 00000 00000 00000 00000	00000 00000 00000 00000 00000	00000 00000 00000 00000 00000	00000 00000 00000 00000 00000	00000 00000 00000 00000 00000	00000 00000 00000 00000 00000	00000 00000 00000 00000 00000	Total ____
	DIARRHOEA	00000 00000	00000 00000	00000 00000	00000 00000	00000 00000	00000 00000	00000 00000	
	MEASLES	00000	00000	00000	00000	00000	00000	00000	
	'AT RISK'	00000 00000	00000 00000	00000 00000	00000 00000	00000 00000	00000 00000	00000 00000	
M O T H E R S	RAISED BLOOD PRESSURE	00000	00000	00000	00000	00000	00000	00000	
	ANAEMIA	00000	00000	00000	00000	00000	00000	00000	
	FIRST ATTENDANCE IN THIRD TRIMESTER	00000	00000	00000	00000	00000	00000	00000	
	'AT RISK'	00000	00000	00000	00000	00000	00000	00000	

Figure 6.5 Tally sheets for data collection for use by village health workers in Tanzania

health status of the user communities. This is often the case where accessibility and availability of care is substandard. Community-based health workers should be trained in simple techniques of data collection and analysis for local use. The introduction of community health registers may be one way of introducing the concept and practice of data gathering at the community level. Such data could include information about births and if possible, including birth weight and head size, deaths and cause of death with age, incidence of vaccine preventable infections, diarrhoea. Such information is useful for the monitoring of health events and for decision-making by the local health committees.

WHO IS TO OBTAIN THE FEEDBACK INFORMATION?

Each type of person who obtains feedback information has advantages and problems. In the past much health care evaluation was done by outside consultants who had the skills and no vested interest or commitment. Moreover, often they hardly knew the country at all, although in time some became much in tune with a particular country's needs and wishes. Nationals from another part of the country may feel just as strange as foreigners in some areas and this can cause problems.

Policy makers and planners often evaluate health care. Their problem is that they are asked to criticise their own plans and sometimes this can be difficult. However, they often do have a good overview. Programme staff and managers working on a particular project also find it difficult to be objective in evaluation. Although they may know very well what defects and deficiencies exist, they may be unable to write about them in case they lose their jobs.

Members of the community may seem to have fewer vested interests in a project yet in many countries certain factions of a community dominate health care programmes. The question is 'Who in the community should evaluate the programme and how can one section of the community be compared with another?' Academics and research staff now do health care evaluation in most countries. They can cause problems for health care personnel because their approach is sometimes too theoretical and academic and often they have little experience of working in remote areas. They may take too long to produce their reports and may not be willing or able to take decisions on what needs to be done. Another risk is the question of whether people doing evaluation should be specialists and if specialised, in what discipline. Although the choice is wide, it is usually easy to decide who is to do the evaluation provided it is clear why evaluation is needed, and what the subject is (see table 6.10).

WHERE SHOULD THE FEEDBACK BE DONE?

What area? What population? What sub-groups?
These are crucial questions in many Districts where there is a diversity of people, health problems, and health care provision. The biases that can occur in visiting only villages near the tarmac road and talking only to élite male informants are well known and have already been referred to.

WHEN SHOULD FEEDBACK INFORMATION BE OBTAINED?

Which season of the year is selected for an evaluation to be done will influence the pattern of disease found, how effectively health personnel are able to do their work, whether the community is very busy from dawn to dusk, and therefore absent from their homes, or available to talk in a slightly less busy time. Many evaluations have been done in a short cool or dry season ignoring the fact that a country may be hot and wet for most of the time. Whatever period is chosen for an evaluation will influence what is found.

Table 6.10 **Who is to carry out the evaluation?**

Type of person	Advantages	Disadvantages
Outside consultants	No vested interests. Will teach new skills to local workers	Sometimes do not know local situation well
Policy makers and planners	Good overview	Difficult to criticise their own plans
Programme staff and managers	Intimate knowledge of the local situation and of working with the programme	May be unwilling to be critical
Community leaders	Project often has the stated aim of serving the community	Leaders may not be truly representative
Poor groups in the community	Project aims to serve the poor	Other groups in the community may obstruct or influence them
Academics and research staff	Usually skilled in the techniques and willing to teach	May take too long to produce the report and raise too many questions
Those working in the project	Immediate feedback. Strengthens the workers' sense of responsibility for their own projects. A natural part of the management process of setting targets, implementation, supervision and control	Requires effective managers and an appropriate managerial system

CONSTRAINTS ON GETTING FEEDBACK

The main constraints usually felt in evaluation are shortages of time, money and staff. A further major problem may occur if the evaluation is not wanted by staff or community or if the information is thought to be wanted for political reasons. If an external agency funds an evaluation it may well impose constraints. The final constraint is that evaluation always requires some effort. People will lose interest in evaluation if they do not see that it leads to changes, or confirms and encourages their current efforts.

FURTHER READING

Fisher A. *Handbook for Family Planning Operations Research Design.* The Population Council, New York, 1991.

Feuerstein M-T. *Partners in Evaluation. Evaluating Development and Community Programmes with Participants.* Macmillan Press Ltd, Basingstoke, 1986.

World Health Organization. *Development of Indicators for Monitoring Progress Towards Health for All by the Year 2000.* (Health for All Series No. 4), WHO, Geneva, 1981.

World Health Organization. *Health Programme Evaluation. Guiding Principles.* (Health for All Series No. 6), WHO, Geneva, 1981.

7 Future Prospects: Challenges for Change

The World Health Organization has identified 'Health For All by the Year 2000' (HFA-2000) as a major objective. Progress towards this target on a global scale has been remarkable. Access to Primary Health Care services which include vaccination, safe water supply and excreta disposal, availability of essential drugs, attendance by a trained person during pregnancy and childbirth and child surveillance has increased dramatically in the past decade. Seventy per cent of the population in the least developed countries now enjoy access to neighbourhood health services compared to less than 20 per cent in the 1970s. Provision of clean water in the rural areas has increased to 68 per cent, and similar progress has also been made with regard to daily energy consumption.

The progress towards HFA-2000 received a further boost from the World Summit on children where the following targets were agreed for the 1990s:

(1) Reducing the under-fives' child mortality by one third or to 70 per thousand live births whichever is the greater reduction.
(2) Halving the maternal mortality rate of 1990.
(3) Halving the 1990 prevalence of severe and moderate malnutrition among children less than five years old.
(4) Providing safe drinking water and sanitary means of disposing of excreta for all.
(5) Ensuring that access to basic education is universal and that 80 per cent of primary school age children of both sexes complete their primary schooling.
(6) Halving, at least, the adult illiteracy rates with stress on female literacy.
(7) Protecting children in especially difficult circumstances such as street children, and victims of war and disaster.

That these targets were agreed by pragmatic political leaders is an indication of how far health care systems have evolved in the developing world since independence in the 1960s. From a nucleus of a predominantly hospital-based elitist system established primarily to serve the purposes of a colonial administration, the health systems of countries have grown to provide a broad coverage first through the basic health care strategy of the 1960s, and then the Primary Health Care approach since 1978.

Along the way there have been debates and controversies. Very early on clarification was needed to establish the distinction between Primary Care and Primary Health Care. It is now generally agreed that whereas Primary Care is first contact care, Primary Health Care refers to the more fundamental services needed for health (see figure 6.1). The second debate was about 'selective' and 'comprehensive' Primary Health Care. Selective care is targeted and therefore, effective in the short term. It is provided by specialist professionals (who do not relinquish power), and more attractive to funding bodies (who are looking for short term success). Even though selective PHC may be made available, accessible and acceptable it leaves out the issues of equity and empowerment of communities for dealing with their endemic health problems. Its proponents have argued that global programmes like oral rehydration therapy (ORT) for diarrhoea, the expanded programme of immunisation (EPI), the Safe Motherhood Initiative and the Child Survival Revolution are, in effect, forms of selective PHC. The truth of the matter is that these programmes are part of the overall approach in comprehensive PHC and represent its cutting edge (see figure 7.1).

Popular expectations all over the world are rising. Many have read reports of miracle drugs and heard about new techniques in surgery and in medical practice. The inability of their governments to provide such facilities in the country is taken as admission of failure. Only a few have heard about Primary Health Care, or about the progress in their own country with regard to immunisation coverage, diarrhoea deaths being prevented and facilities being created for assisting women during childbirth. Hospital specialists and consultants who dominate the health care systems in most countries have tended to obstruct the shift of emphasis in national health plans from hospital-based services to Primary Health Care within the community. Since they also control medical education, changes in undergraduate and postgraduate medical training have been minimal. For maintaining progress towards HFA-2000 the greatest need is to train the District level medical and nursing officers of the future. On their shoulders will rest the responsibility of providing the administrative, logistical, managerial and technical skills needed for PHC. If the need for training such health workers within the country is accepted, then even a more urgent need is for training the future teachers of such cadres.

Experience with PHC has shown that the District hospital has the important role of being the springboard for PHC in the District. In order to do so a variety of new functions and activities have to be added starting at Level C (District hospital), but then spreading to Levels B (Health Centres) and A (the community) as experience is gained. These activities may be considered in a modular form as shown in figure 7.2. This has been increasingly the

Future prospects

VILLAGE HEALTH COMMITTEE
(Provides health volunteers, birth attendants, storage and clinic facilities, other resources. Also mobilises community and gives social support)

ADULT LITERACY GROUPS
(Functional literacy)

YOUTH ACTIVITIES
(Leisure activities, literacy and handicraft classes)

FARMERS' CLUB
(Agricultural improvement; backyard gardening; poultry and veterinary classes)

PARENTS' CLUBS
(Nutrition of children; hygiene; immunisation; counselling of young parents)

A core programme of health development

- Under-fives' clinics
- Antenatal and Maternity Care
- Nutrition intervention activities
- Communicable disease control including mass campaigns
- Family health counselling
- Curative care
- Procurement of supplies and essentials
- Diarrhoeal disease control including water resources development, sanitation and oral rehydration

DAY CARE and CRECHE
(Child care and stimulation; support of breastfeeding; supervision during weaning)

WOMEN'S CLUBS
(Sewing and cooking classes; health and family life education)

SEWING and HANDICRAFT GROUPS
(Income generation activities; preparation of weaning foods)

NUTRITION REHABILITATION ACTIVITIES
(Cooking demonstration; use of multi-mixes; agricultural techniques)

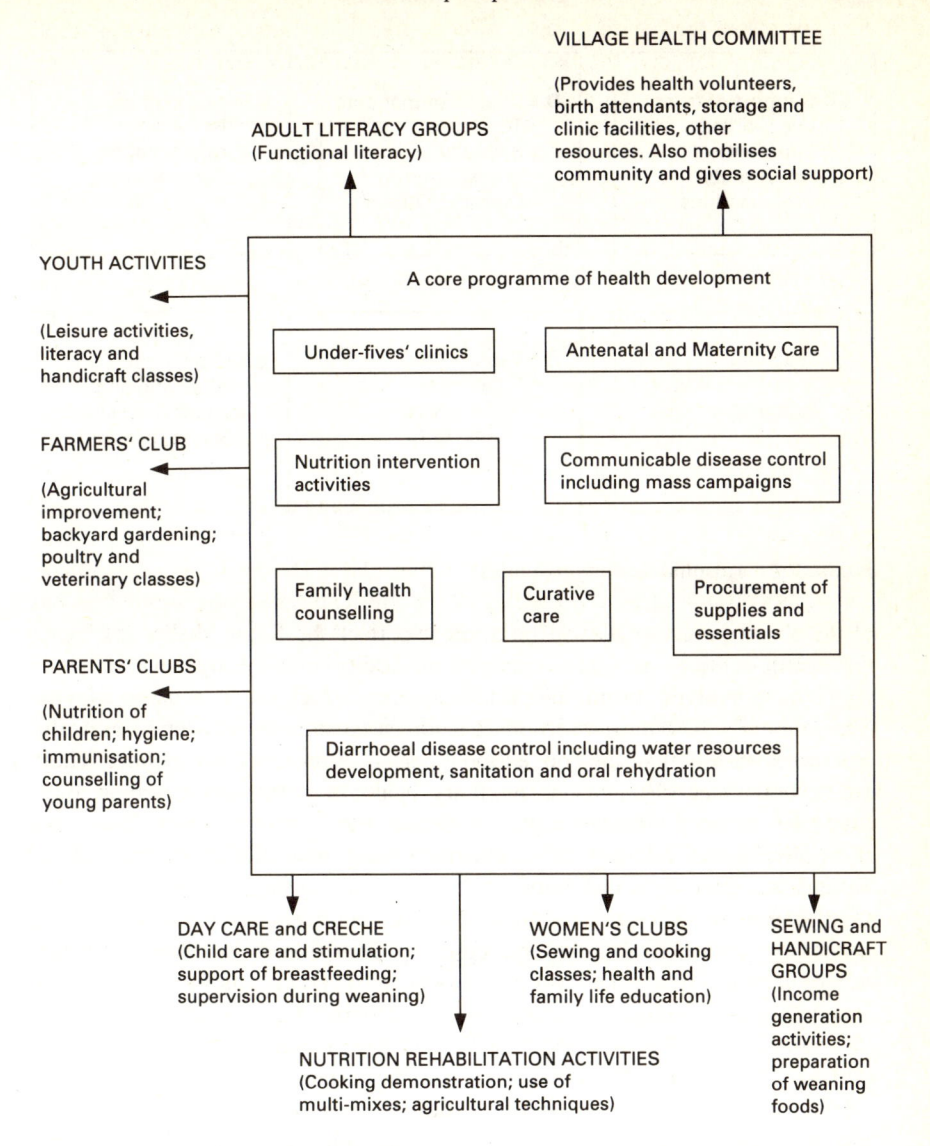

Figure 7.1 A core health programme

pattern of development in the 1990s, though inevitably countries differ in the extent to which ramification of activities has been achieved at Levels A and B.

These programmes need to be integrated into the national strategy for PHC and may well provide the springboard for other activities in Primary Health Care. The administrative, logistical, technical and managerial needs

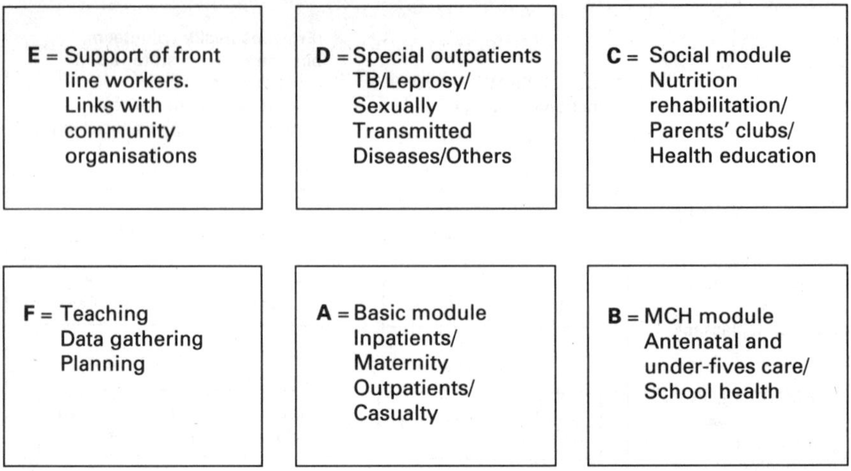

E = Support of front line workers. Links with community organisations	**D** = Special outpatients TB/Leprosy/ Sexually Transmitted Diseases/Others	**C** = Social module Nutrition rehabilitation/ Parents' clubs/ Health education
F = Teaching Data gathering Planning	**A** = Basic module Inpatients/ Maternity Outpatients/ Casualty	**B** = MCH module Antenatal and under-fives care/ School health

Figure 7.2 Hospital activity modules

of the above programmes will be a major part of the future challenges facing the health services of most countries. In addition, there are several trends which have evolved during the past decade and which are now being adopted internationally. The most widespread amongst these is the utilisation of health workers with less than the conventional period of professional training. There are several categories of such auxiliary health workers and different countries have adopted different standards depending upon their needs. They vary from physicians' assistants, med-exes and nurse practitioners to medical assistants, rural medical aides and maternal and child health aides. Moreover, semi-literate or illiterate community health volunteers are now increasingly becoming part of the national strategies for provision of health care, so that village health workers (VHW) and trained birth attendants (TBA) are to be found in many countries. In the average District of a developing country there are likely to be between 40 to 80 auxiliary health workers and between 100 to 200 VHWs and TBAs. They are being considered as the front-line workers or the grass-roots of the health system by national planners. The task of providing them with logistical and technical support is immense and will require considerable managerial skills. Besides, they need to be carefully integrated into the chain of health facilities and the network of health care. Their training needs are also likely to be many and varied, and require careful study. Any service is completely dependent upon the quality of its personnel. A sound policy for development of health manpower is necessary to deal with morale at all levels.

When a District hospital achieves the goal of spreading PHC activities at all the Level B and A service points, then it has in essence achieved an equitable distribution of health resources for the District population.

Countries like Sri Lanka, Costa Rica and Thailand who have led the way have achieved this objective through community involvement for exercising social control over resources at all levels. This is one way of doing away with elitism in health care (see figure 7.3).

Other trends and innovations are all linked to the above trend of utilising less trained health workers. Because of their short training and background of education, their armamentarium of drugs has to be limited, and restricted to the common illnesses. Partly on account of this and partly due to economic reasons, many countries have drawn up a list of essential drugs. The managers of the District Health Teams will be continuously faced with the task of deciding between getting essential supplies to the front-line workers and procuring sophisticated equipment or expensive medicines for the hospital consultants. If hospitals are allowed to deprive front-line workers of their basic tools, then PHC will exist only on paper, and some demarcation of funding is therefore necessary. Thirdly, in the drive to simplify medical care many standardised techniques have evolved like the arm band for measuring arm circumferences, multi-mixes as weaning foods, the cereal-based oral rehydration solution and so on. Such new developments need to be communicated to health workers at all levels of care and so systems for communication of new ideas will have to be evolved. Programmes of distance teaching for the front-line workers in order to provide continual in-service training may need to be considered as part of the managerial process.

Health service reforms are taking place in the more developed countries, with transfer of resources (and also costs) from hospitals to the community level. Economic necessity together with movements of populations and changes in demographic patterns have led to bed and ward closures. These reforms have thrown up a number of issues about health service management in all countries.

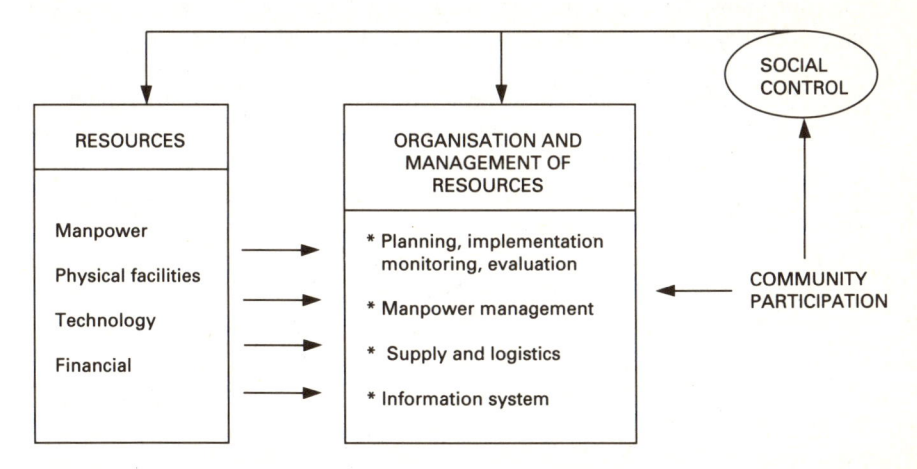

Figure 7.3 Social control over health resources management

MANAGEMENT ISSUES LIKELY TO ARISE DURING THE NEXT DECADE

Four major management issues are apparent from the analysis of plans for Primary Health Care. These are: the need for support of teams at Level B (the Health Station); the need to solve the already existing Level B problems so that they are not perpetuated into Primary Health Care (Level A); the need to find alternative approaches to problems of accessibility and coverage; and the need to set priorities for local community (Level A) health tasks to be done.

It is obvious from the detailed analysis of workload and responsibilities that the Health Centre and sub-centre (Level B) teams are being asked to become a major pivot point in the proposed PHC system. They will need to be recognised as such in this important role and considerable support is likely to be needed for their work. Training, supplies, supervision and morale boosts are needed. These have not been forthcoming. Change in medical and nursing curricula is a lengthy process of academic bargaining between departments. Undergraduate training continues to remain disease oriented rather than preparation for delivering Primary Health Care to communities. Even if the curricula were to change, training manuals and texts are not being produced locally, but imported from abroad and adapted for the local situation. For most developing countries training in the professional skills needed for PHC is still occurring abroad at great expense. Thus, government initiatives for orienting the health services to PHC remain starved of academic inputs. The most urgent need is for transfer of training to the developing world.

Unless the existing problems at Level B are solved they are likely to be perpetuated into the new Primary Health Care system. New community work commenced before facing up to and solving outstanding current rural health problems will very soon be ineffective. If, for example, a midwife at Level B does not know the importance of recognising high risk pregnancies she will have difficulty in supervising referral of such high risk pregnancies to her by the traditional birth attendant. Clearly, once such a problem is recognised, it means that midwives already in the post need to be given refresher training first, before the TBA training programme starts. But will such health care problems at present existing in the health care system be recognised? Will action be taken to solve them or will they be extended further into the system and perpetuated when PHC activities are initiated?

Calculations of expected accessibility by communities to the components of the Primary Health Care system have been made. The limiting factor is rightly recognised as travelling time from the home to the source of health care. A local community service is needed within 15 minutes of travel for everyone. Where walking is the common mode of transport such a service needs to be within one mile. The next more sophisticated level of service needs to be within one hour's travel. If walking, this should be not more than 5 miles (8 km) away. Hospital services need to be within 4–5 hours travel (to enable, for example, an emergency caesarian section in good time for an obstructed labour). This will need to be within 25 miles (40 km) assuming that emergencies and very ill patients will be carried by vehicular transport. This, of course, makes no allowance for being caught in traffic jams. Clearly, as transport methods change health service accessibility changes. Although an optimistic view suggests that mechanised transport could become available to more people through good public transport systems, more taxis or an increased number of private cars, it is also probable that in many countries fewer people will be able to travel by motorised vehicle or by animal transport because the cost has become too high.

With varying accessibility, health workers have varying degrees of responsibility for populations 'covered' by their services. Health workers need to have a specified catchment area that they are serving; yet as population movements change, the population coverage also needs to change with it.

Another problem with accessibility is that in sparsely-populated areas there may be relatively few people living within one mile (1.6 km) radius (Level A), or five-mile (8 km) radius (Level B). In the same country there may be a very large population in a one mile (1.6 km) radius in the squatter area of the capital city and very few in a mountainous region. If communities with population densities of less than 200 people in a one-mile (1.6 km) catchment area are excluded, whole sections of the poorest parts of a country are likely to be left out. How are health services to be provided in these sparsely populated areas? Clearly different options are needed. In a very densely populated area it may be efficient to have several 'specialised' workers, for example, in mother and child health care, environmental health, community development and so on. In a sparsely-populated area, it is far more likely that a multi-purpose worker will be needed who can do the more important tasks in each of these four specialities. National broadcasting stations can be persuaded to broadcast health programmes at specified times with multi-purpose workers in remote communities acting as local discussion leaders. Such an approach has been successfully tried in Lesotho and Tanzania.

Overall, recognition of these four major management issues draws attention to the options available for making progress towards health for all by the year 2000. These include:

(1) Aiming to provide a network of local community (Level A) health development posts and supportive services as well as training from the district hospitals as outlined in the plan in chapter 3.
(2) Aiming to spread Level B Health Centre/sub-centre services by extending existing services so that units are available at five mile (8 km) intervals and at the same time solving existing managerial or administrative problems as well as encouraging outreach services. Such services may also serve as means for commencing a dialogue with local communities.

(3) Aiming to strengthen Level C elements of the strategy by emphasising support and logistic services; integrated intersectoral approaches and in-service training; co-ordination with the District Council; and team work between the doctors, public health staff, health administration and environmental health, with regular planning, monitoring and evaluation of health activities.
(4) Focusing on communicable disease control, Mother and Child Health services and environmental hygiene at every level.
(5) Aiming to teach specified Level A tasks to school teachers; for example, rehydration for diarrhoea, malaria prophylaxis, measuring malnutrition, accident prevention and so on.
(6) Aiming to make the more important among Level A skills as part of the general knowledge of everyone during the coming decade through adult education, school education and the use of mass media.
(7) A combination of any of the above.

OBSTACLES AND CONSTRAINTS

Several difficulties which may impede progress towards health for all by the year 2000 also need to be recognised. The most important of these is the question of a national will. The creation of a national will for rural health care and upliftment is not a matter of issuing directives or decrees, but one of a national dialogue at all levels. And this kind of awakening is still lacking in many countries. Then there is the perennial problem of finance. Heavy burden of international debt coupled with inflation inevitably leads to dependence on loans from the World Bank and the International Monetary Fund. The consequent imposition of structural adjustment policies means cutbacks in the social services like health, education and welfare. The overall effect of structural adjustment has been an increase in poverty and all the health and social problems which stem from it. Resources in real terms may be diminishing at a time when more resources are needed. Political problems may also rear their heads. Confronted with all these difficulties, how real is the government commitment to health for all? Will lip-service, political rhetoric and grandiose policy be backed up by real resources and action? Will there be changes in curricula to provide appropriate training, for doctors and nurses as well as other types of health workers? Will career structures reflect this new training? Within the health care system will special roles and experiences be well utilised? For example, will those who have worked in isolated places for many years to help patients with leprosy, tuberculosis or mental illness be included as trainers? Is there really any hope of improving procurement and distribution of dressings and drugs? At the local level, will the local political structure be a constraint to community development? Will the special problem of disparities in the cities be recognised, for example, the shanty towns with only token services on the one hand and the larger hospitals rapidly consuming health care resources on the other? Will the social distance between most health workers and their communities remain so great that true 'community involvement' is impossible as health workers fail to comprehend and respect the local culture in their area?

New health problems have arisen with tragic consequences and requiring urgent attention at a time when the health services are under great strain. Infection with the HIV virus is continuing to decimate communities with a heavy social and economic toll. Substance abuse and alcoholism are both related to the epidemic and are all part of a feeling of hopelessness and demoralisation.

Economic recession has led to the growth of monetarism and the market approach in health planning. Several leading nations, for example Britain, have introduced health service reforms based on these principles. Many previously colonial countries are presented with a dilemma. Their health systems were based on those of the ruling country including training, organisation and ways of funding. Secondly, in bringing about reforms, politicians tend to interpret key concepts of PHC according to political expediency and dogma. The administrative and managerial structures which are emerging in metropolitan countries may not necessarily be appropriate for the previous dependencies. Under monetarist ideology the District Health Authority is no longer the monopolist supplier of health care. Instead, it is mainly to act as a purchaser who must look for the best buys for the population it represents. Hospitals and practitioners, including the private sector, become independent suppliers. Political patronage and the proliferation of administrators can be a major drain on resources. These and similar other developments do present new challenges with regard to methods of community involvement in health programmes and maintaining standards of care. New questions to be asked are:

(1) Are there important organisations, or groups of practitioners providing health care besides the local Ministry of Health services?
(2) Are there formal channels which District Health Managers can use to encourage them to participate in health care delivery?
(3) How are they to be financed?
(4) What should be the mechanisms of monitoring and control?

OPPORTUNITIES FOR FUTURE GROWTH

There are potential problems but there is also considerable hope. Under-used resources exist as well as unsolved problems (see table 7.1).

Already many international agencies (for example, the World Bank) are looking on health as a productive asset not a cost. It is likely that government

Table 7.1 **Problems and opportunities for future plans**

Problems	Opportunities
Distribution, logistics, communications	Community-based distribution utilising the successful methods of commercial organisations and businesses
Poor use of health personnel	Updating training and continuous education, distance teaching
Community health workers seem to need teaching the most elementary topics	Improve the basic education of everyone via primary schools, mass media, child-to-child activities
Health care seems to be a continuous, ever-growing burden to society.	Take more action on the social causes of ill health, poverty, environmental hazards, etc.
Lack of community health personnel	Training and career structure needed for doctors, nurses and others; integration of traditional care providers like traditional healers, birth attendants etc. into the health care system

bureaucracies will come to recognise this point of view and health will receive higher priority in national expenditure.

Never before has there been so much opportunity to share health knowledge so quickly with people around the world. The information technology revolution is only just beginning. The possibilities of improving training programmes, running frequent refresher courses, enabling continuous education and providing primary and functional adult literacy for everyone are no longer remote. Satellites, radio, television, and listening groups using learning packages, have the potential for helping people to take better care of themselves, to look after their surroundings and to remove hazards to health, and to be able to make better use of the health care system.

The disease care system in all countries complains of lack of resources and of difficulties in obtaining modern equipment and drugs. But perhaps the real challenge is to use available resources better. In many countries the large majority spend part of their wealth (in cash or as gifts) to purchase treatment when they are ill. Often this is from the local herbalist or traditional practitioner, or the nearby drug store or chemist. When this expenditure has been studied (for example, in Botswana) it has been found to be as much as the government itself spends on disease care. Could this money be used more effectively to improve health? Could people spend more effectively on preventing ill health, by improving water supply, using safer machinery and so on? And maybe this is linked to the inadequate distribution network. If aerated drinks and baby foods can find their way to every corner of a country, why not essential drugs?

Health services in developing countries have hitherto been mainly concerned with physical disease. With greater understanding of the epidemiology of illness in the community, the high prevalence of mental illness is gradually being recognised in all countries. In many of the industrial societies where community mental health services are better organised it is estimated that one in eight amongst males and one in ten amongst females is at risk of developing mental illness at some time during life. Similar data for the developing countries are not available. But social disparities, poverty and polarisation within the society tend to be greater in the developing nations and the incidence of mental ill-health is likely to be as high as in the industrial world, if not higher. Promotion of mental health through a network of services will be a major challenge in many countries.

Another important potential growth point lies at the heart of the new concepts about health. Although many people still equate health care with disease care, some are already recognising that health is not just the absence of disease. Health is to do with adapting, it is a 'freedom from the dark seas of disease', stemming from spiritual and social well-being as well as physical health. There is a Zulu saying that a person is a reflection of the people around him. As social malaise becomes commonplace, as violence, intimidation and fear rack cities in the world, more and more 'health' is likely to be seen in terms of social well-being as well as individual physical and emotional fitness.

Changes in the health system, new hazards like the HIV epidemic, and developments within the management science itself have brought about several new trends in health management. Basically, management in health, as indeed elsewhere, is about managing self, managing resources, and managing others. Early developments in management science were the results of insight and convergence between management on the one hand and social and behavioural sciences on the other. A similar convergence between management and economics is occurring now because of the need for better management of resources. The next development is for management and epidemiological sciences to come together in order to quantify disease and the etiologic factors. Such a convergence will also be valuable for evaluating health care, and for measuring the effects of different interventions. This calls for designing appropriate training programmes in epidemiologic methods and tools. This is the current trend in many countries.

THE UNTAPPED RESOURCES

Within the existing health care system one under-used potential lies in the nurses and in paramedical staff. Many have been under-utilised and underrated for too long. Some nurses are already being given better training and appropriate career structures to fulfil their potential. Public health nurses in particular now have seven years of excellent training. In many countries they are being used very effectively to bridge community and hospital work. Physiotherapists are now being recognised as key people to help the old and disabled live as fully as possible within their community. Their in-depth training in anatomy and physiology as well as in physical therapy and aids gives them a potential to be key members of the District Health Team.

Perhaps there are personnel in the community who are under-utilised too; grandmothers in all societies are a main source of education in parenthood and practical child care; families in many societies still care for their elderly, the sick and the chronic sick or the terminally ill. Health care in the future can choose to build on these roles. Many societies have herbal knowledge, food preferences and taboos, and ways of looking after the sick in body or spirit. Much care within the family is never seen by the formal health care system. But when societies change and the informal system breaks down with migration or other disruptions, the formal health care system is faced with taking over such traditional roles of the family. Hence it is advisable to carefully nurture the caring and providing role of the family and strengthen it with services like home visiting and District nursing. But even when the health system does have to take on the caring and supportive functions traditionally provided by the family, there is scope for linkage. Volunteers may well come forward to help. Often they are just those wise women and helpers in the neighbourhood who in traditional society performed similar functions and roles.

BEYOND PRIMARY HEALTH CARE?

There are always options and there will always be change. One type of Primary Health Care is certainly no panacea for all health and health care problems. As health problems change and communities become older, and if they show the increasing frequencies of chronic and degenerative disease found in the currently developed world, new patterns of health care will be needed in developing countries. They may well include the same central elements of Primary Health Care but a system of care focused more on the old, the handicapped and chronic sick, and not, as at present, on mothers and children, the acutely ill and on prevention of communicable diseases. There will be new health care problems influenced by the political and economic situation prevailing at the time, by new training for health workers and by their perception of their roles. But if the health care system is flexible enough to monitor and respond to the changes in disease as well as changes in resources and political commitment, there could well be good health beyond the year 2000.

Index

Accessibility 96
Age group
  for priority, 34–35
  with great need 34–36
Assessing,
  health centre data 43–46
  health needs 33–83
  team effectiveness 194
  vital statistics 43
At-risk concept 60–61

Budgeting 114–116
Buildings 248–253

Catchment area 70–71
Change
  challenges for 281–292
  force field analysis 219–220
  implementation of 164, 217–223
  levers of 166–167
  management of 163–167
  monitoring of 218
  organisational 163–167
  planning for 215
  recognising the need for 162, 215
Child care
  need for 34, 129
Clustering 46–47
Communication 208, 212
Community
  health management problems 3
  involvement 18
  operational research 81–82
  round 79
  surveys 79–81
Community assessment
  methods of 78–82
Community diagnosis
  for evaluation and feedback 260–261
  of health care problems 56
  of health care resources 64
  of ill health 34
  of priority problems and
    tasks to be done 95
Conflicts
  management of 223–227
Community round 79

Community involvement 102–104
Constraints 288–289
Counselling staff 241
Coverage with health services 57–58
Cycle of planning and implementation 89–114

Deaths
  common causes of 35, 42–45
Delegation 187–189
Demographic factors 34
Determinants of disease 52–56
District Health Plan 84–139
  health care systems 86
  five pillars of 86
  weaknesses in 85–86
  key concepts in 93–114
  key elements 156
District level problems 2
District health system 140–143
District health team 18–19
District Primary Health Care
  epidemiology of 34–56
  identifying health needs 33–82
  requirements of a programme 13–23
Drugs and vaccines
  management of 252–253
Decision-making 27
Doctors as managers 26–27, 31

Environmental health 129–130
Environmental influence 53–56
Epidemiological profile 34–56
  disease priorities 35–46
  distribution of problems 38–42
  priority age groups 34–35
  seasonal influences 48
Evaluation 63, 256–279
  choice of methods 272–277
  data collection forms 271–273
  elements of PHC for 263–264
  five basic questions 256
  levels of 261
  methods of obtaining 267–269
  monitoring 258–260
  pitfalls in 261
  potential of 265
  questions to be answered 269
  reasons for 256–268
  selection of time, place and person 277–279
  specific questions for consideration 266–268

subject for 260–261
types of 257

Failure of health improvement 9–13
  additional problems 12
  dilemma in health care 10
  unbalanced spending 11
Family health specialist
  responsibilities of 131–132
Family health workers
  training guidelines for 125
Feedback
  choice of methods 268–270
  levels of 266–267
  methods and procedures 267–269
  pitfalls in 261
  reasons for 256–268
  specific questions for consideration 267
  subject for 263
  types of 257
Finance 242–248
  operating budgets 244
Financial resources 75–77
Force Field Analysis 219–220
Formal organisation 144–145
Functional manpower planning 100–101
Future prospects 281–292
  issues for next decade 286–288
  obstacles 288–289
  opportunities 289–291
  untapped resources 291–292

Group effectiveness
  assessment of 194
Group work 195–197

Health care
  demographic factors 34
  need for new approach 8–13
  new management requirements 13–23
  problems with services 2–4
  significant new developments 15
  specific questions for evaluation 267
Health care system 96
  functions and task definitions in 96–99
Health Centre
  managerial problems 3
Health for all
  obstacles and constraints 288–289
  opportunities for future growth 289–291
  options for progress 287–288

Health management
  issues in 1, 8
Health manpower planning 68–70
  training 125–128
Health needs
  assessment of 42–43
  determining of 42–43
  epidemiological data 46–48
  key questions in 33
Health organisation 140–167
  as a multi-level structure 20–21
  as a network 146
  as a skill pyramid 146
  as a system 140–144, 146–150
  defective, symptoms of 162–163
  formal and informal 144–146
  job description 154
  key elements of 155–157
  key principles of 151–155
  management of 157–161
  organisational change 163–167
  organisational culture 151
Health problems 38
  clustering of 46–47
Health resources
  distribution of 5–13
  distribution of doctors 6
  identification of 64–78
  urban-rural disparity 9–11
Health services
  coverage 57–58
  delivery problems 80–82
  evaluation 63
  levels of accessibility 96
  MCH care 58–59
  pattern of organisation 190
  priority problems and functions of 95
  problems with 56–64
  quality 61–62
  staff interaction 62–63
  standards 178–180
  use of at-risk concept 60–61
  utilisation of 59–60
Hospital
  managerial problems 3–4

Job description 154

Key concepts in planning 93–114
Key result areas 172–173

Leadership 26–29
  key abilities of 27–29

Liaison 105–107
Local resources 64–78

Management
  by objectives 171–176
  communication in 208–212
  delegation 187–189
  district health team 18–20
  framework for 169–170
  in health 8–13
  in local socio-cultural environment 160–161
  issues for the next decade 286–288
  motivation in 199–208
  new approaches in 13–19
  of building 248–252
  of change 212–223
  of conflicts 223–227
  of drugs, vaccines and pharmaceuticals 252–253
  of delegation 187–189
  of finance 242–248
  of meetings 197–199
  of motivation 199–207
  of personnel 229–241
  of Primary Health Care 16–19
  of standards and discipline 178–180, 236–241
  of supervisors 227–229
  of supplies and stores 250–252
  of systems 170
  of training 232–234
  of transport 253–255
  participatory 176–178
  personal skills in 183–255
  poor, symptoms of 8
  practical 168–255
  requirements for Primary Health Care 16–23
  role of doctors 31
  roles of managers 171
  staff deployment 12
  symptoms of poor management 8
  teamwork in 189–195
  theory X and Y 202–203
  types of 109
Management style 30–31
Matching resources to needs 105
Material and labour 74–75
Maternity care
  requirements for 35–41
Maternity services 117–121
Maternal mortality 36–37
MCH care 58–59

Medical auxiliaries 72–74
Meetings 197–199
Microplanning 63–64
Monitoring 258–260
Motivation 199–208
  theories of 201–205
Multi-level health system 22

Natural resources 77–78
Needs and resources 105
New approaches in health 13–23
  need for 4
Non-professional as health resource 72–74
Nutritional problems 52–54
  interaction with infection 52

Objectives
  key result areas 172–173
  management by 171–176
Obstetric care 39–41
Operational research 81–82
Organisational change 163–167
Organisational culture 151

Parasitic diseases 54
Personnel
  management of 229–234
Planning
  advantages and pitfalls of 89–93, 122–129
  as a learning process 93
  calculating staff requirement 130–133
  causes of failure 92, 122–123
  dangers of 89–92
  district health 84–139
  examples of 117–121
  implementation 168–255
  key concepts in 93–114
  liaison and teamwork 105–107
  linking tasks with community involvement 102–104
  matching needs and resources 105
  of training procedures 105
  prevention of failure 92
  problems exacerbating difficulties 12–13
  process of 89–114
  reasons for failure of 122–124
  recognising current defects 118
  task definition 100–102
  task setting based on priorities 98–99

Plan 86–114
Planning process 89–114
   advantages and pitfalls 89–93, 122–124
   checklist 115
   key concepts in 93–114
Primary Health Care
   community involvement 18
   differences in management of 20
   district health team 18–20
   essential elements for evaluation 263–265
   features of 13–14, 16
   global response to 24–26
   management requirements of 14–23
   multi-level structure 20–21
   of the future 292
   requirements for a district programme 157–161
   untapped resources 291–292
Primary health worker
   profile of 23
Priority health problems
   and tasks to be done 95, 98–99
   task analysis 97
Problems
   at community level 3
   at District level 2
   at Health Centre level 3
   in hospitals 3
   of maternity services 117–121
   overall 4
   with health services 2
Process of planning 89–114
   advantages and pitfalls 89–93, 122–124
   as learning method 93
Project proposals 132–139

Quality of health care 61–62

Rapid epidemiological assessment 48–52
Resource allocation 114–116

Safe motherhood 38
School children as health resource 74
Seasonal influence 48
Shared vision 28
Skill pyramid 146–147
Snowballing 94
Social malaise indicators 48
Social marketing 104

Sources of health care 64–70
   catchment area 70–71
   medical auxiliaries 72–74
   planning for 67–68
   traditional practitioners 64–66
Staff development 234–236
   role of training in 232–233
Staff
   interaction of 62–63
   requirement 128–132
Standards 178–180
   and discipline
   of safety 180–182
Supervisors
   job of 227–228
   managing support for 228–229
Supplies and stores 250–252
Surveys 79–82
System of health organisation 22–23, 146–149, 170
   examples of 148
   features of 149
   modern 150
   traditional 151

Target setting 109
Task analysis 100–102
   definition of 99–100
   linking with community involvement 102–104
Teaching methods 126
Team work 105–107, 189–195
Time
   management of 183–186
Traditional healers 65–66
Training 232
   and staff development 234–236
   for team work 110–114
   guidelines for family health workers 125
   key procedures in 110–114
   management of 232
   pitfalls in 127–128
   planning for 124–127
   regular updating 112
   task oriented 110
   trainers of community health workers 125–126
Transport
   management of 253–255

Urban-rural disparity 9–11
Utilisation of services 59–60